About the [illegible]

Titu Andreescu received his BA, MS, and PhD from the West University of Timisoara, Romania. The topic of his doctoral dissertation was "Research on Diophantine Analysis and Applications." He has served as director of the MAA American Mathematics Competitions (1998-2003), coach of the USA International Mathematical Olympiad Team (IMO) for 10 years (1993-2002), director of the Mathematical Olympiad Summer Program (1995-2002) and leader of the USA IMO Team (1995-2002). In 2002 Titu was elected member of the IMO Advisory Board, the governing body of the international competition. Titu received the Edyth May Sliffe Award for Distinguished High School Mathematics Teaching from the MAA in 1994 and a "Certificate of Appreciation" from the president of the MAA in 1995 for his outstanding service as coach of the Mathematical Olympiad Summer Program in preparing the US team for its perfect performance in Hong Kong at the 1994 International Mathematical Olympiad.

Zuming Feng graduated with a PhD from Johns Hopkins University with emphasis on Algebraic Number Theory and Elliptic Curves. He teaches at Phillips Exeter Academy. He also served as a coach of the USA IMO team (1997-2003), the deputy leader of the USA IMO Team (2000-2002), and an assistant director of the USA Mathematical Olympiad Summer Program (1999-2002). He is a member of the USA Mathematical Olympiad Committee since 1999, and has been the leader of the USA IMO team and the academic director of the USA Mathematical Olympiad Summer Program since 2003. He received the Edyth May Sliffe Award for Distinguished High School Mathematics Teaching from the MAA in 1996 and 2002.

About the Authors

Titu Andreescu received his BA, MS, and PhD from the West University of Timisoara, Romania. The topic of his doctoral dissertation was "Research on Diophantine Analysis and Applications". Titu served as director of the MAA American Mathematics Competitions (1998-2003), coach of the USA International Mathematical Olympiad Team (IMO) for 10 years (1993-2002), director of the Mathematical Olympiad Summer Program (1995-2002) and leader of the USA IMO Team (1995-2002). In 2002 Titu was elected member of the IMO Advisory Board, the governing body of the international competition. Titu received the Edyth May Sliffe Award for Distinguished High School Mathematics Teaching from the MAA in 1994 and a "Certificate of Appreciation" from the president of the MAA in 1995 for his outstanding service as coach of the Mathematical Olympiad Summer Program in preparing the US team for its perfect performance in Hong Kong at the 1994 International Mathematical Olympiad.

Zuming Feng graduated with a PhD from Johns Hopkins University with emphasis on Algebraic Number Theory and Elliptic Curves. He teaches at Phillips Exeter Academy. He also served as a coach of the USA IMO team (1997-2003), the deputy leader of the USA IMO Team (2000-2002), and an assistant director of the USA Mathematical Olympiad Summer Program (1999-2002). He is a member of the USA Mathematical Olympiad Committee since 1999, and has been the leader of the USA IMO team and the academic director of the USA Mathematical Olympiad Summer Program since 2003. He received the Edyth May Sliffe Award for Distinguished High School Mathematics Teaching from the MAA in 1996 and 2002.

Titu Andreescu
Zuming Feng

103 Trigonometry Problems

From the Training of the USA IMO Team

Birkhäuser
Boston • Basel • Berlin

Titu Andreescu
University of Wisconsin
Department of Mathematical
and Computer Sciences
Whitewater, WI 53190
U.S.A.

Zuming Feng
Phillips Exeter Academy
Department of Mathematics
Exeter, NH 03833
U.S.A.

AMS Subject Classifications: Primary: 97U40, 00A05, 00A07, 51-XX; Secondary: 11L03, 26D05, 33B10, 42A05

Library of Congress Cataloging-in-Publication Data
Andreescu, Titu, 1956-
103 trigonometry problems : from the training of the USA IMO team / Titu Andreescu, Zuming Feng.
p. cm.
Includes bibliographical references.
ISBN 0-8176-4334-6 (acid-free paper)
1. Trigonometry–Problems, exercises, etc. I. Title: One hundred and three trigonometry problems. II. Feng, Zuming. III. Title.

QA537.A63 2004
516.24–dc22 2004045073

ISBN 0-8176-4334-6 Printed on acid-free paper.

 Birkhäuser 

Printed in the United States of America.

9 8 7 6 5 4 3 2 1 SPIN 10982723

www.birkhauser.com

Contents

Preface

This book contains 103 highly selected problems used in the training and testing of the U.S. International Mathematical Olympiad (IMO) team. It is not a collection of very difficult, impenetrable questions. Instead, the book gradually builds students' trigonometric skills and techniques. The first chapter provides a comprehensive introduction to trigonometric functions, their relations and functional properties, and their applications in the Euclidean plane and solid geometry. This chapter can serve as a textbook for a course in trigonometry. This work aims to broaden students' view of mathematics and better prepare them for possible participation in various mathematical competitions. It provides in-depth enrichment in important areas of trigonometry by reorganizing and enhancing problem-solving tactics and strategies. The book further stimulates interest for the future study of mathematics.

In the United States of America, the selection process leading to participation in the International Mathematical Olympiad (IMO) consists of a series of national contests called the American Mathematics Contest 10 (AMC 10), the American Mathematics Contest 12 (AMC 12), the American Invitational Mathematics Examination (AIME), and the United States of America Mathematical Olympiad (USAMO). Participation in the AIME and the USAMO is by invitation only, based on performance in the preceding exams of the sequence. The Mathematical Olympiad Summer Program (MOSP) is a four-week intensive training program for approximately 50 very promising students who have risen to the top in the American Mathematics Competitions. The six students representing the United States of America in the IMO are selected on the basis of their USAMO scores and further testing that takes place during MOSP.

Throughout MOSP, full days of classes and extensive problem sets give students thorough preparation in several important areas of mathematics. These topics include combinatorial arguments and identities, generating functions, graph theory, recursive relations, sums and products, probability, number theory, polynomials, functional equations, complex numbers in geometry, algorithmic proofs, combinatorial and advanced geometry, functional equations, and classical inequalities.

Olympiad-style exams consist of several challenging essay problems. Correct solutions often require deep analysis and careful argument. Olympiad questions can seem impenetrable to the novice, yet most can be solved with elementary high school mathematics techniques, cleverly applied.

Here is some advice for students who attempt the problems that follow.

- Take your time! Very few contestants can solve all the given problems.

- Try to make connections between problems. An important theme of this work is that all important techniques and ideas featured in the book appear more than once!

- Olympiad problems don't "crack" immediately. Be patient. Try different approaches. Experiment with simple cases. In some cases, working backwards from the desired result is helpful.

- Even if you can solve a problem, do read the solutions. They may contain some ideas that did not occur in your solutions, and they may discuss strategic and tactical approaches that can be used elsewhere. The solutions are also models of elegant presentation that you should emulate, but they often obscure the tortuous process of investigation, false starts, inspiration, and attention to detail that led to them. When you read the solutions, try to reconstruct the thinking that went into them. Ask yourself, "What were the key ideas? How can I apply these ideas further?"

- Go back to the original problem later, and see whether you can solve it in a different way. Many of the problems have multiple solutions, but not all are outlined here.

- Meaningful problem-solving takes practice. Don't get discouraged if you have trouble at first. For additional practice, use the books on the reading list.

Acknowledgments

Thanks to Dorin Andrica and Avanti Athreya, who helped proofread the original manuscript. Dorin provided acute mathematical ideas that improved the flavor of this book, while Avanti made important contributions to the final structure of the book. Thanks to David Kramer, who copyedited the second draft. He made a number of corrections and improvements. Thanks to Po-Ling Loh, Yingyu Gao, and Kenne Hon, who helped proofread the later versions of the manuscript.

Many of the ideas of the first chapter are inspired by the Math 2 and Math 3 teaching materials from the Phillips Exeter Academy. We give our deepest appreciation to the authors of the materials, especially to Richard Parris and Szczesny "Jerzy" Kaminski.

Many problems are either inspired by or adapted from mathematical contests in different countries and from the following journals:

- *High-School Mathematics*, China
- *Revista Matematică Timişoara*, Romania

We did our best to cite all the original sources of the problems in the solution section. We express our deepest appreciation to the original proposers of the problems.

Abbreviations and Notation

Abbreviations

AHSME	American High School Mathematics Examination
AIME	American Invitational Mathematics Examination
AMC10	American Mathematics Contest 10
AMC12	American Mathematics Contest 12, which replaces AHSME
APMC	Austrian–Polish Mathematics Competition
ARML	American Regional Mathematics League
IMO	International Mathematical Olympiad
USAMO	United States of America Mathematical Olympiad
MOSP	Mathematical Olympiad Summer Program
Putnam	The William Lowell Putnam Mathematical Competition
St. Petersburg	St. Petersburg (Leningrad) Mathematical Olympiad

Notation for Numerical Sets and Fields

$\mathbb{Z}$	the set of integers
$\mathbb{Z}_n$	the set of integers modulo n
$\mathbb{N}$	the set of positive integers
$\mathbb{N}_0$	the set of nonnegative integers
$\mathbb{Q}$	the set of rational numbers
$\mathbb{Q}^+$	the set of positive rational numbers
$\mathbb{Q}^0$	the set of nonnegative rational numbers
$\mathbb{Q}^n$	the set of n-tuples of rational numbers
$\mathbb{R}$	the set of real numbers
$\mathbb{R}^+$	the set of positive real numbers
$\mathbb{R}^0$	the set of nonnegative real numbers
$\mathbb{R}^n$	the set of n-tuples of real numbers
$\mathbb{C}$	the set of complex numbers
$[x^n](p(x))$	the coefficient of the term x^n in the polynomial $p(x)$

Notation for Sets, Logic, and Geometry

$\|A\|$	the number of elements in the set A
$A \subset B$	A is a proper subset of B
$A \subseteq B$	A is a subset of B
$A \setminus B$	A without B (set difference)
$A \cap B$	the intersection of sets A and B
$A \cup B$	the union of sets A and B
$a \in A$	the element a belongs to the set A
a, b, c	lengths of sides BC, CA, AB of triangle ABC
A, B, C	angles $\angle CAB$, $\angle ABC$, $\angle BCA$ of triangle ABC
R, r	circumradius and inradius of triangle ABC
$[\mathcal{F}]$	area of region $\mathcal{F}$
$[ABC]$	area of triangle ABC
$\|BC\|$	length of line segment BC
$\widehat{AB}$	the arc of a circle between points A and B

103 Trigonometry Problems

1
Trigonometric Fundamentals

Definitions of Trigonometric Functions in Terms of Right Triangles

Let S and T be two sets. A **function** (or **mapping** or **map**) f from S to T (written as $f : S \to T$) assigns to each $s \in S$ exactly one element $t \in T$ (written $f(s) = t$); t is the **image** of s. For $S' \subseteq S$, let $f(S')$ (the image of S') denote the set of images of $s \in S'$ under f. The set S is called the **domain** of f, and $f(S)$ is the **range** of f.

For an angle θ (Greek "theta") between $0°$ and $90°$, we define trigonometric functions to describe the size of the angle. Let rays OA and OB form angle θ (see Figure 1.1). Choose point P on ray OA. Let Q be the **foot** (that is, the bottom) of the perpendicular line segment from P to the ray OB. Then we define the sine (sin), cosine (cos), tangent (tan), cotangent (cot), cosecant (csc), and secant (sec) functions as follows, where $|PQ|$ denotes the length of the line segment PQ:

$$\sin\theta = \frac{|PQ|}{|OP|}, \quad \csc\theta = \frac{|OP|}{|PQ|},$$
$$\cos\theta = \frac{|OQ|}{|OP|}, \quad \sec\theta = \frac{|OP|}{|OQ|},$$
$$\tan\theta = \frac{|PQ|}{|OQ|}, \quad \cot\theta = \frac{|OQ|}{|PQ|}.$$

First we need to show that these functions are well defined; that is, they only depends on the size of θ, but not the choice of P. Let P_1 be another point lying on ray OA, and let Q_1 be the foot of perpendicular from P_1 to ray OB. (By the way, "P sub 1" is how P_1 is usually read.) Then it is clear that right triangles OPQ and OP_1Q_1 are similar, and hence pairs of corresponding ratios, such as $\frac{|PQ|}{|OP|}$ and $\frac{|P_1Q_1|}{|OP_1|}$, are all equal. Therefore, all the trigonometric functions are indeed well defined.

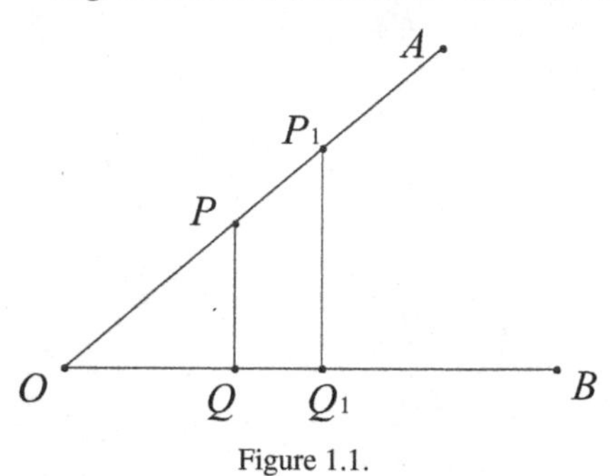

Figure 1.1.

By the above definitions, it is not difficult to see that $\sin\theta$, $\cos\theta$, and $\tan\theta$ are the reciprocals of $\csc\theta$, $\sec\theta$, and $\cot\theta$, respectively. Hence for most purposes, it is enough to consider $\sin\theta$, $\cos\theta$, and $\tan\theta$. It is also not difficult to see that

$$\frac{\sin\theta}{\cos\theta} = \tan\theta \quad \text{and} \quad \frac{\cos\theta}{\sin\theta} = \cot\theta.$$

By convention, in triangle ABC, we let a, b, c denote the lengths of sides BC, CA, and AB, and let $\angle A$, $\angle B$, and $\angle C$ denote the angles CAB, ABC, and BCA. Now, consider a right triangle ABC with $\angle C = 90°$ (Figure 1.2).

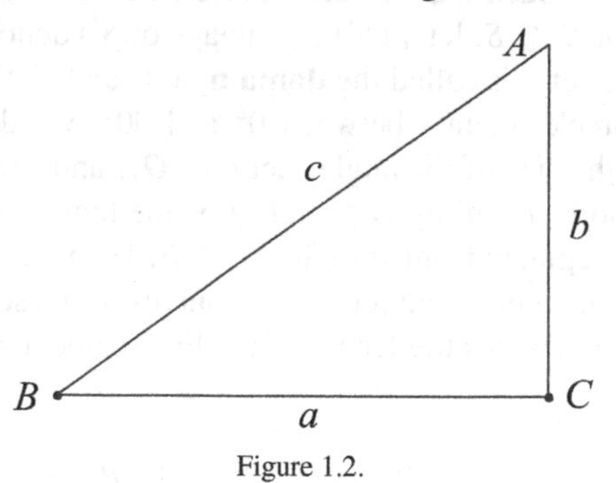

Figure 1.2.

For abbreviation, we write $\sin A$ for $\sin\angle A$. We have

$$\sin A = \frac{a}{c}, \qquad \cos A = \frac{b}{c}, \qquad \tan A = \frac{a}{b};$$

$$\sin B = \frac{b}{c}, \qquad \cos B = \frac{a}{c}, \qquad \tan B = \frac{b}{a};$$

and

$$\begin{array}{llll} a = c\sin A, & a = c\cos B, & a = b\tan A; & \\ b = c\sin B, & b = c\cos A, & b = a\tan B; & \\ c = a\csc A, & c = a\sec B, & c = b\csc B, & c = b\sec A. \end{array}$$

It is then not difficult to see that if A and B are two angles with $0^\circ < A, B < 90^\circ$ and $A+B = 90^\circ$, then $\sin A = \cos B$, $\cos A = \sin B$, $\tan A = \cot B$, and $\cot A = \tan B$. In the right triangle ABC, we have $a^2 + b^2 = c^2$. It follows that

$$(\sin A)^2 + (\cos A)^2 = \frac{a^2}{c^2} + \frac{b^2}{c^2} = 1.$$

It can be confusing to write $(\sin A)^2$ as $\sin A^2$. (Why?) For abbreviation, we write $(\sin A)^2$ as $\sin^2 A$. We have shown that for $0^\circ < A < 90^\circ$,

$$\sin^2 A + \cos^2 A = 1.$$

Dividing both sides of the above equation by $\sin^2 A$ gives

$$1 + \cot^2 A = \csc^2 A, \quad \text{or} \quad \csc^2 A - \cot^2 A = 1.$$

Similarly, we can obtain

$$\tan^2 A + 1 = \sec^2 A, \quad \text{or} \quad \sec^2 A - \tan^2 A = 1.$$

Now we consider a few special angles.

In triangle ABC, suppose $\angle A = \angle B = 45^\circ$, and hence $|AC| = |BC|$ (Figure 1.3, left). Then $c^2 = a^2 + b^2 = 2a^2$, and so $\sin 45^\circ = \sin A = \frac{a}{c} = \frac{1}{\sqrt{2}} = \frac{\sqrt{2}}{2}$. Likewise, we have $\cos 45^\circ = \frac{\sqrt{2}}{2}$ and $\tan 45^\circ = \cot 45^\circ = 1$.

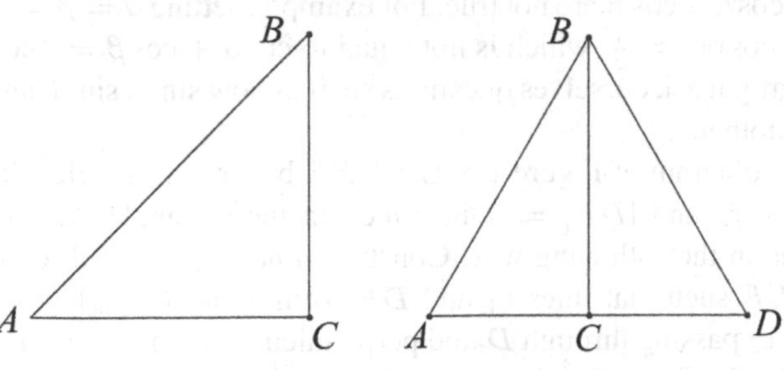

Figure 1.3.

In triangle ABC, suppose $\angle A = 60^\circ$ and $\angle B = 30^\circ$ (Figure 1.3, right). We reflect A across line BC to point D. By symmetry, $\angle D = 60^\circ$, so triangle ABD is equilateral. Hence, $|AD| = |AB|$ and $|AC| = \frac{|AD|}{2}$. Because ABC is a right triangle, $|AB|^2 = |AC|^2 + |BC|^2$. So we have $|BC|^2 = |AB|^2 - \frac{|AB|^2}{4} = \frac{3|AB|^2}{4}$, or $|BC| = \frac{\sqrt{3}|AB|}{2}$. It follows that $\sin 60^\circ = \cos 30^\circ = \frac{\sqrt{3}}{2}$, $\sin 30^\circ = \cos 60^\circ = \frac{1}{2}$, $\tan 30^\circ = \cot 60^\circ = \frac{\sqrt{3}}{3}$, and $\tan 60^\circ = \cot 30^\circ = \sqrt{3}$.

We provide one exercise for the reader to practice with right-triangle trigonometric functions. In triangle ABC (see Figure 1.4), $\angle BCA = 90^\circ$, and D is the foot of the perpendicular line segment from C to segment AB. Given that $|AB| = x$ and $\angle A = \theta$, express all the lengths of the segments in Figure 1.4 in terms of x and θ.

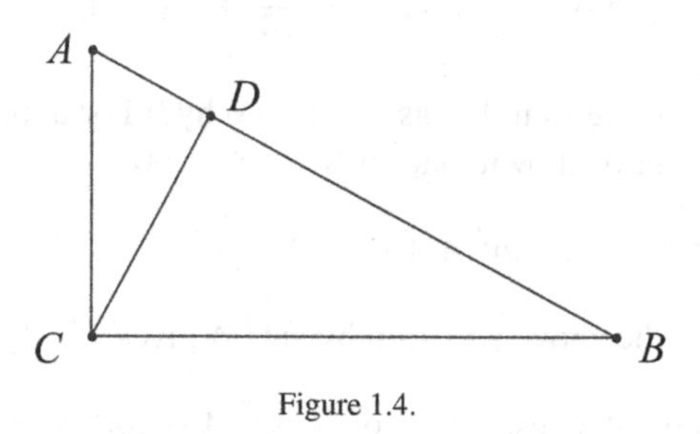

Figure 1.4.

Think Within the Box

For two angles α (Greek "alpha") and β (Greek "beta") with $0^\circ < \alpha, \beta, \alpha + \beta < 90^\circ$, it is not difficult to note that the trigonometric functions do not satisfy the additive distributive law; that is, identities such as $\sin(\alpha + \beta) = \sin\alpha + \sin\beta$ and $\cos(\alpha+\beta) = \cos\alpha + \cos\beta$ are not true. For example, setting $\alpha = \beta = 30^\circ$, we have $\cos(\alpha + \beta) = \cos 60^\circ = \frac{1}{2}$, which is not equal to $\cos\alpha + \cos\beta = 2\cos 30^\circ = \sqrt{3}$. Naturally, we might ask ourselves questions such as how $\sin\alpha$, $\sin\beta$, and $\sin(\alpha+\beta)$ relate to one another.

Consider the diagram of Figure 1.5. Let DEF be a right triangle with $\angle DEF = 90^\circ$, $\angle FDE = \beta$, and $|DF| = 1$ inscribed in the rectangle $ABCD$. (This can always be done in the following way. Construct line ℓ_1 passing through D outside of triangle DEF such that lines ℓ_1 and DE form an acute angle congruent to α. Construct line ℓ_2 passing through D and perpendicular to line ℓ_1. Then A is the foot of the perpendicular from E to line ℓ_1, and C the foot of the perpendicular from F to ℓ_2. Point B is the intersection of lines AE and CF.)

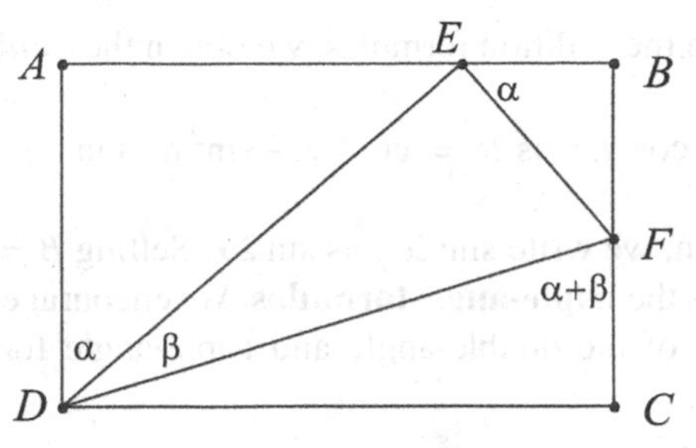

Figure 1.5.

We compute the lengths of the segments inside this rectangle. In triangle DEF, we have $|DE| = |DF| \cdot \cos\beta = \cos\beta$ and $|EF| = |DF| \cdot \sin\beta = \sin\beta$. In triangle ADE, $|AD| = |DE| \cdot \cos\alpha = \cos\alpha\cos\beta$ and $|AE| = |DE| \cdot \sin\alpha = \sin\alpha\cos\beta$. Because $\angle DEF = 90°$, it follows that $\angle AED + \angle BEF = 90° = \angle AED + \angle ADE$, and so $\angle BEF = \angle ADE = \alpha$. (Alternatively, one may observe that right triangles ADE and BEF are similar to each other.) In triangle BEF, we have $|BE| = |EF| \cdot \cos\alpha = \cos\alpha\sin\beta$ and $|BF| = |EF| \cdot \sin\alpha = \sin\alpha\sin\beta$. Since $AD \parallel BC$, $\angle DFC = \angle ADF = \alpha + \beta$. In right triangle CDF, $|CD| = |DF| \cdot \sin(\alpha+\beta) = \sin(\alpha+\beta)$ and $|CF| = |DF| \cdot \cos(\alpha+\beta) = \cos(\alpha+\beta)$.

From the above, we conclude that

$$\cos\alpha\cos\beta = |AD| = |BC| = |BF| + |FC| = \sin\alpha\sin\beta + \cos(\alpha+\beta),$$

implying that

$$\cos(\alpha+\beta) = \cos\alpha\cos\beta - \sin\alpha\sin\beta.$$

Similarly, we have

$$\sin(\alpha+\beta) = |CD| = |AB| = |AE| + |EB| = \sin\alpha\cos\beta + \cos\alpha\sin\beta;$$

that is,

$$\sin(\alpha+\beta) = \sin\alpha\cos\beta + \cos\alpha\sin\beta.$$

By the definition of the tangent function, we obtain

$$\begin{aligned}\tan(\alpha+\beta) &= \frac{\sin(\alpha+\beta)}{\cos(\alpha+\beta)} = \frac{\sin\alpha\cos\beta + \cos\alpha\sin\beta}{\cos\alpha\cos\beta - \sin\alpha\sin\beta}\\ &= \frac{\frac{\sin\alpha}{\cos\alpha} + \frac{\sin\beta}{\cos\beta}}{1 - \frac{\sin\alpha\sin\beta}{\cos\alpha\cos\beta}} = \frac{\tan\alpha + \tan\beta}{1 - \tan\alpha\tan\beta}.\end{aligned}$$

We have thus proven the **addition formulas** for the sine, cosine, and tangent functions for angles in a restricted interval. In a similar way, we can develop an addition formula for the cotangent function. We leave it as an exercise.

By setting $\alpha = \beta$ in the addition formulas, we obtain the **double-angle formulas**

$$\sin 2\alpha = 2\sin\alpha\cos\alpha, \ \cos 2\alpha = \cos^2\alpha - \sin^2\alpha, \ \tan 2\alpha = \frac{2\tan\alpha}{1-\tan^2\alpha},$$

where for abbreviation, we write $\sin(2\alpha)$ as $\sin 2\alpha$. Setting $\beta = 2\alpha$ in the addition formulas then gives us the **triple-angle formulas**. We encourage the reader to derive all the various forms of the double-angle and triple-angle formulas listed in the Glossary of this book.

You've Got the Right Angle

Because of the definitions of the trigonometric functions, it is more convenient to deal with trigonometric functions in the context of right triangles. Here are three examples.

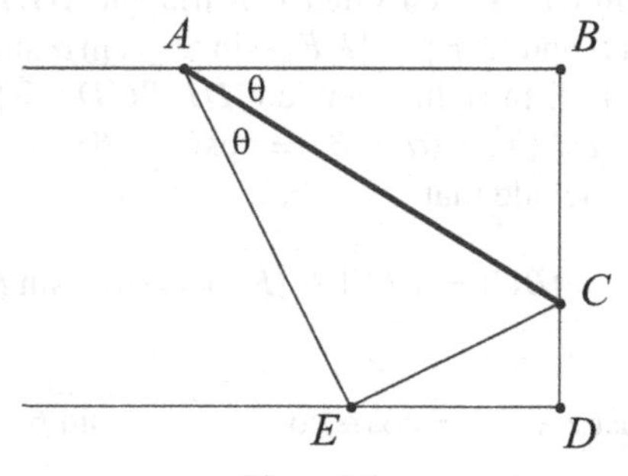

Figure 1.6.

Example 1.1. Figure 1.6 shows a long rectangular strip of paper, one corner of which has been folded over along AC to meet the opposite edge, thereby creating angle θ ($\angle CAB$ in Figure 1.6). Given that the width of the strip is w inches, express the length of the crease AC in terms of w and θ. (We assume that θ is between $0°$ and $45°$, so the real folding situation is consistent with the configuration shown in Figure 1.6.)

We present two solutions.

First Solution: In the right triangle ABC, we have $|BC| = |AC|\sin\theta$. In the right triangle AEC, we have $|CE| = |AC|\sin\theta$. (Indeed, by folding, triangles ABC and AEC are congruent.) Because $\angle BCA = \angle ECA = 90° - \theta$, it follows that $\angle BCE = 180° - 2\theta$ and $\angle DCE = 2\theta$ (Figure 1.7). Then, in the right triangle CDE, $|CD| = |CE|\cos 2\theta$. Putting the above together, we have

$$w = |BD| = |BC| + |CD| = |AC|\sin\theta + |AC|\sin\theta\cos 2\theta,$$

implying that

$$|AC| = \frac{w}{\sin\theta(1+\cos 2\theta)}.$$

■

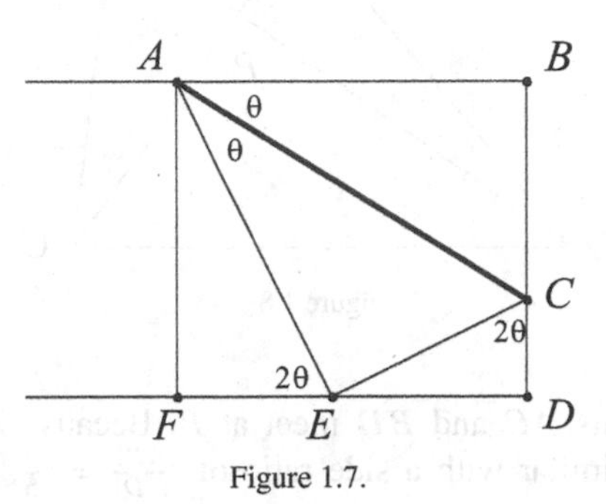

Figure 1.7.

Second Solution: Let F be the foot of the perpendicular line segment from A to the opposite edge. Then in the right triangle AEF, $\angle AEF = 2\theta$ and $|AF| = w$. Thus $|AF| = |AE|\sin 2\theta$, or $|AE| = \frac{w}{\sin 2\theta}$. In the right triangle AEC, $\angle CAE = \angle CAB = \theta$ and $|AE| = |AC|\cos\theta$. Consequently,

$$|AC| = \frac{|AE|}{\cos\theta} = \frac{w}{\sin 2\theta\cos\theta}.$$

■

Putting these two approaches together, we have

$$|AC| = \frac{w}{\sin\theta(1+\cos 2\theta)} = \frac{w}{\sin 2\theta\cos\theta},$$

or $\sin\theta(1+\cos 2\theta) = \sin 2\theta\cos\theta$. Interested readers can use the formulas we developed earlier to prove this identity.

Example 1.2. In the trapezoid $ABCD$ (Figure 1.8), $AB \parallel CD$, $|AB| = 4$ and $|CD| = 10$. Suppose that lines AC and BD intersect at right angles, and that lines BC and DA, when extended to point Q, form an angle of 45°. Compute $[ABCD]$, the area of trapezoid $ABCD$.

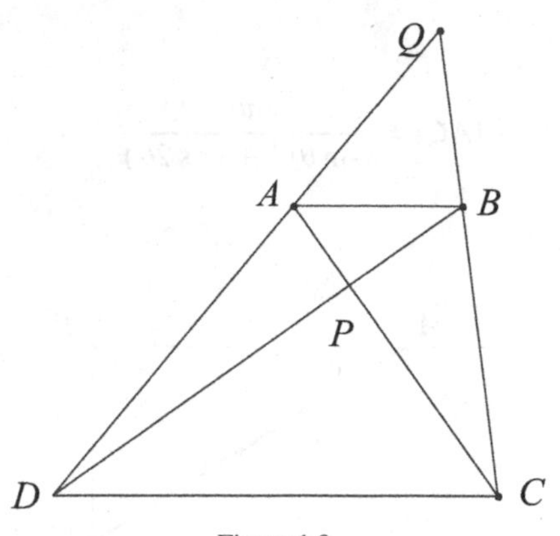

Figure 1.8.

Solution: Let segments AC and BD meet at P. Because $AB \parallel CD$, triangles ABP and CDP are similar with a side ratio of $\frac{|AB|}{|CD|} = \frac{2}{5}$. Set $|AP| = 2x$ and $|BP| = 2y$. Then $|CP| = 5x$ and $|DP| = 5y$. Because $\angle APB = 90^\circ$, $[ABCD] = \frac{1}{2}|AC| \cdot |BD| = \frac{49xy}{2}$. (To see this, consider the following calculation: $[ABCD] = [ABD] + [CBD] = \frac{1}{2}|AP| \cdot |BD| + \frac{1}{2}|CP| \cdot |BD| = \frac{1}{2}|AC| \cdot |BD|$.)

Let $\alpha = \angle ADP$ and $\beta = \angle BCP$. In right triangles ADP and BCP, we have

$$\tan\alpha = \frac{|AP|}{|DP|} = \frac{2x}{5y} \quad \text{and} \quad \tan\beta = \frac{|BP|}{|CP|} = \frac{2y}{5x}.$$

Note that $\angle CPD = \angle CQD + \angle QCP + \angle QDP$, implying that $\alpha + \beta = \angle QCP + \angle QDP = 45^\circ$. By the **addition formulas**, we obtain that

$$1 = \tan 45^\circ = \tan(\alpha + \beta) = \frac{\tan\alpha + \tan\beta}{1 - \tan\alpha\tan\beta} = \frac{\frac{2x}{5y} + \frac{2y}{5x}}{1 - \frac{2x}{5y}\frac{2y}{5x}} = \frac{10(x^2 + y^2)}{21xy},$$

which establishes that $xy = \frac{10(x^2+y^2)}{21}$. In triangle ABP, we have $|AB|^2 = |AP|^2 + |BP|^2$, or $16 = 4(x^2 + y^2)$. Hence $x^2 + y^2 = 4$, and so $xy = \frac{40}{21}$. Consequently,

$$[ABCD] = \frac{49xy}{2} = \frac{49}{2} \cdot \frac{40}{21} = \frac{140}{3}.$$

■

Example 1.3. [AMC12 2004] In triangle ABC, $|AB| = |AC|$ (Figure 1.9). Points D and E lie on ray BC such that $|BD| = |DC|$ and $|BE| > |CE|$. Suppose that $\tan\angle EAC$, $\tan\angle EAD$, and $\tan\angle EAB$ form a geometric progression, and that $\cot\angle DAE$, $\cot\angle CAE$, and $\cot\angle DAB$ form an arithmetic progression. If $|AE| = 10$, compute $[ABC]$, the area of triangle ABC.

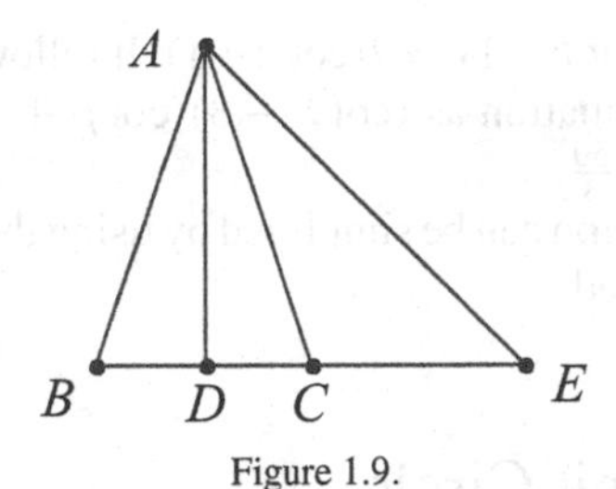

Figure 1.9.

Solution: We consider right triangles ABD, ACD, and ADE. Set $\alpha = \angle EAD$ and $\beta = \angle BAD = \angle DAC$. Then $\angle EAC = \alpha - \beta$ and $\angle EAB = \alpha + \beta$. Because $\tan \angle EAC$, $\tan \angle EAD$, and $\tan \angle EAB$ form a geometric progression, it follows that

$$\tan^2 \alpha = \tan^2 \angle EAD = \tan \angle EAC \tan \angle EAB = \tan(\alpha - \beta)\tan(\alpha + \beta).$$

By the **addition formulas**, we obtain

$$\tan^2 \alpha = \frac{\tan \alpha + \tan \beta}{1 - \tan \alpha \tan \beta} \cdot \frac{\tan \alpha - \tan \beta}{1 + \tan \alpha \tan \beta} = \frac{\tan^2 \alpha - \tan^2 \beta}{1 - \tan^2 \alpha \tan^2 \beta},$$

or

$$\tan^2 \alpha - \tan^4 \alpha \tan^2 \beta = \tan^2 \alpha - \tan^2 \beta.$$

Hence, $\tan^4 \alpha \tan^2 \beta = \tan^2 \beta$, and so $\tan \alpha = 1$, or $\alpha = 45^\circ$. (We used the fact that both $\tan \alpha$ and $\tan \beta$ are positive, because $0^\circ < \alpha, \beta < 90^\circ$.) Thus ADE is an isosceles right triangle with $|AD| = |DE| = \frac{|AE|}{\sqrt{2}} = 5\sqrt{2}$. In the right triangle ACD, $|DC| = |AD| \tan \beta$, and so $[ABC] = |AD| \cdot |CD| = |AD|^2 \tan \beta = 50 \tan \beta$.

Because $\cot \angle DAE = \cot 45^\circ = 1$, $\cot \angle CAE$, and $\cot \angle DAB$ form an arithmetic progression, it follows that

$$2\cot(45^\circ - \beta) = 2\cot \angle CAE = \cot \angle DAE + \cot \angle DAB = 1 + \cot \beta.$$

Setting $45^\circ - \beta = \gamma$ (Greek "gamma") in the above equation gives $2\cot \gamma = 1 + \cot \beta$. Because $0^\circ < \beta, \gamma < 45^\circ$, applying the addition formulas gives

$$1 = \cot 45^\circ = \cot(\beta + \gamma) = \frac{\cot \beta \cot \gamma - 1}{\cot \beta + \cot \gamma},$$

or $\cot \beta + \cot \gamma = \cot \beta \cot \gamma - 1$. Solving the system of equations

$$\begin{cases} 2\cot \gamma = 1 + \cot \beta, \\ \cot \beta + \cot \gamma = \cot \beta \cot \gamma - 1 \end{cases} \quad \text{or} \quad \begin{cases} 2\cot \gamma = \cot \beta + 1, \\ \cot \gamma(\cot \beta - 1) = \cot \beta + 1 \end{cases}$$

for $\cot\beta$ gives $(\cot\beta+1)(\cot\beta-1) = 2(\cot\beta+1)$. It follows that $\cot^2\beta - 2\cot\beta - 3 = 0$. Factoring the last equation as $(\cot\beta - 3)(\cot\beta + 1) = 0$ gives $\cot\beta = 3$. Thus $[ABC] = 50\tan\beta = \frac{50}{3}$. ■

Of course, the above solution can be simplified by using the **subtraction formulas**, which will soon be developed.

Think Along the Unit Circle

Let ω denote the unit circle, that is, the circle of radius 1 centered at the origin $O = (0, 0)$. Let A be a point on ω in the first quadrant, and let θ denote the acute angle formed by line OA and the x axis (Figure 1.10). Let A_1 be the foot of the perpendicular line segment from A to the x axis. Then in the right triangle AA_1O, $|OA| = 1$, $|AA_1| = \sin\theta$, and $|OA_1| = \cos\theta$. Hence $A = (\cos\theta, \sin\theta)$.

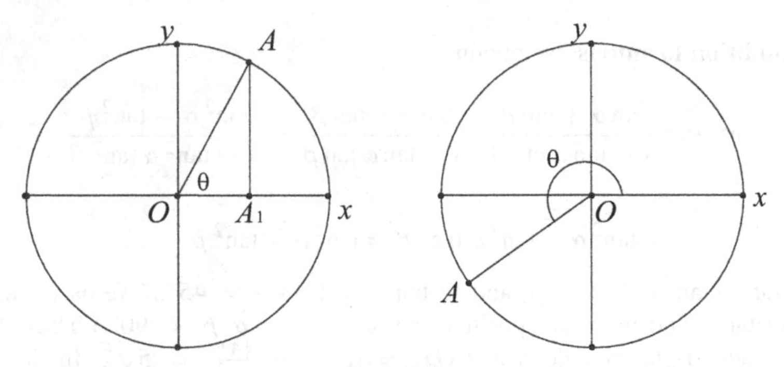

Figure 1.10.

In the coordinate plane, we define a **standard angle** (or **polar angle**) formed by a ray ℓ from the origin and the positive x axis as an angle through which the positive x axis can be rotated to coincide with ray ℓ. Note that we have written *a* standard angle and not *the* standard angle. That is because there are many ways in which the positive x axis can be rotated in order to coincide with the ray ℓ. In particular, a standard angle of $\theta_1 = x^\circ$ is equivalent to a standard angle of $\theta_2 = x^\circ + k \cdot 360^\circ$, for all integers k. For example, a standard angle of 180° is equivalent to all of these standard angles: $\ldots, -900^\circ, -540^\circ, -180^\circ, +540^\circ, +900^\circ, \ldots$. Thus a standard angle is a directed angle. By convention, a positive angle indicates rotation of the x axis in the counterclockwise direction, while a negative standard angle indicates that the x axis is turned in the clockwise direction.

We can also define **the** standard angle formed by two lines in the plane as the smallest angle needed to rotate one line in the counterclockwise direction to coincide with the other line. Note that this angle is always greater than or equal to 0° and less than 180°.

For a point A in the plane, we can also describe the position of A (relative to the origin) by the distance $r = |OA|$ and standard angle θ formed by the line OA and the x axis. These coordinates are called **polar coordinates**, and they are written in the form $A = (r, \theta)$. (Note that the polar coordinates of a point are not unique.)

In general, for any angle θ, we define the values of $\sin\theta$ and $\cos\theta$ as the coordinates of points on the unit circle. Indeed, for any θ, there is a unique point $A = (x_0, y_0)$ (in rectangular coordinates) on the unit circle ω such that $A = (1, \theta)$ (in polar coordinates). We define $\cos\theta = x_0$ and $\sin\theta = y_0$; that is, $A = (\cos\theta, \sin\theta)$ if and only if $A = (1, \theta)$ in polar coordinates.

From the definition of the sine and cosine functions, it is clear that for all integers k, $\sin(\theta + k \cdot 360°) = \sin\theta$ and $\cos(\theta + k \cdot 360°) = \cos\theta$; that is, they are **periodic functions** with **period** 360°. For $\theta \neq (2k + 1) \cdot 90°$, we define $\tan\theta = \frac{\sin\theta}{\cos\theta}$; and for $\theta \neq k \cdot 180°$, we define $\cot\theta = \frac{\cos\theta}{\sin\theta}$. It is not difficult to see that $\tan\theta$ is equal to the slope of a line that forms a standard angle of θ with the x axis.

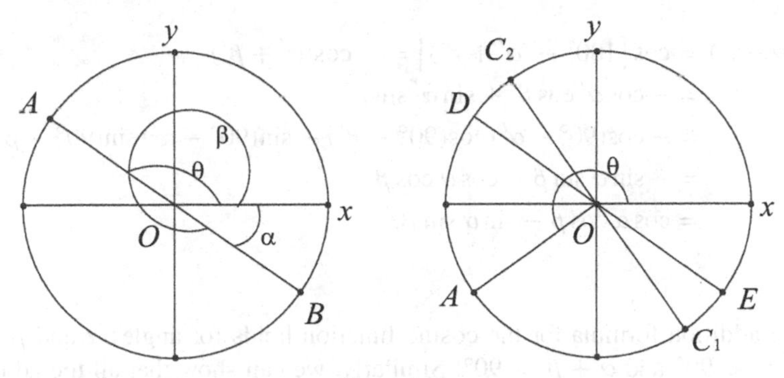

Figure 1.11.

Assume that $A = (\cos\theta, \sin\theta)$. Let B be the point on ω diametrically opposite to A. Then $B = (1, \theta + 180°) = (1, \theta - 180°)$. Because A and B are symmetric with respect to the origin, $B = (-\cos\theta, -\sin\theta)$. Thus

$$\sin(\theta \pm 180°) = -\sin\theta, \quad \cos(\theta \pm 180°) = -\cos\theta.$$

It is then easy to see that both $\tan\theta$ and $\cot\theta$ are functions with a period of 180°. Similarly, by rotating point A around the origin 90° in the counterclockwise direction (to point C_2 in Figure 1.11), in the clockwise direction (to C_1), reflecting across the x axis (to D), and reflecting across the y axis (to E), we can show that

$$\begin{aligned}
\sin(\theta + 90°) &= \cos\theta, & \cos(\theta + 90°) &= -\sin\theta,\\
\sin(\theta - 90°) &= -\cos\theta, & \cos(\theta - 90°) &= \sin\theta,\\
\sin(-\theta) &= -\sin\theta, & \cos(-\theta) &= \cos\theta,\\
\sin(180° - \theta) &= \sin\theta, & \cos(180° - \theta) &= -\cos\theta.
\end{aligned}$$

Furthermore, by either reflecting A across the line $y = x$ or using the second and third formulas above, we can show that $\sin(90° - \theta) = \cos\theta$ and $\cos(90° - \theta) = \sin\theta$. This is the reason behind the nomenclature of the "cosine" function: "cosine" is the complement of sine, because the angles $90° - \theta$ and θ are complementary angles. All these interesting and important trigonometric identities are based on the geometric properties of the unit circle.

Earlier, we found **addition and subtraction formulas** defined for angles α and β with $0° < \alpha, \beta < 90°$ and $\alpha + \beta < 90°$. Under our general definitions of trigonometry functions, we can expand these formulas to hold for all angles. For example, we assume that α and β are two angles with $0° \le \alpha, \beta < 90°$ and $\alpha + \beta > 90°$. We set $\alpha' = 90° - \alpha$ and $\beta' = 90° - \beta$. Then α' and β' are angles between $0°$ and $90°$ with a sum of less than $90°$. By the **addition formulas** we developed earlier, we have

$$\begin{aligned}
\cos(\alpha + \beta) &= \cos\left[180° - (\alpha' + \beta')\right] = -\cos(\alpha' + \beta') \\
&= -\cos\alpha' \cos\beta' + \sin\alpha' \sin\beta' \\
&= -\cos(90° - \alpha')\cos(90° - \beta') + \sin(90° - \alpha')\sin(90° - \beta') \\
&= -\sin\alpha \sin\beta + \cos\alpha \cos\beta \\
&= \cos\alpha \cos\beta - \sin\alpha \sin\beta.
\end{aligned}$$

Thus, the addition formula for the cosine function holds for angles α and β with $0° \le \alpha, \beta < 90°$ and $\alpha + \beta > 90°$. Similarly, we can show that all the addition formulas developed earlier hold for all angles α and β. Furthermore, we can prove the **subtraction formulas**

$$\begin{aligned}
\sin(\alpha - \beta) &= \sin\alpha \cos\beta - \cos\alpha \sin\beta, \\
\cos(\alpha - \beta) &= \cos\alpha \cos\beta + \sin\alpha \sin\beta, \\
\tan(\alpha - \beta) &= \frac{\tan\alpha - \tan\beta}{1 + \tan\alpha \tan\beta}.
\end{aligned}$$

We call these, collectively, the **addition and subtraction formulas**. Various forms of the **double-angle** and **triple-angle** formulas are special cases of the addition and subtraction formulas. Double-angle formulas lead to various forms of the **half-angle formulas**. It is also not difficult to check the **product-to-sum formulas** by the addition and subtraction formulas. We leave this to the reader. For angles α and

β, by the addition and subtraction formulas, we also have

$$\begin{aligned}\sin\alpha+\sin\beta &= \sin\left(\frac{\alpha+\beta}{2}+\frac{\alpha-\beta}{2}\right)+\sin\left(\frac{\alpha+\beta}{2}-\frac{\alpha-\beta}{2}\right)\\ &= \sin\frac{\alpha+\beta}{2}\cos\frac{\alpha-\beta}{2}+\cos\frac{\alpha+\beta}{2}\sin\frac{\alpha-\beta}{2}\\ &\quad+\sin\frac{\alpha+\beta}{2}\cos\frac{\alpha-\beta}{2}-\cos\frac{\alpha+\beta}{2}\sin\frac{\alpha-\beta}{2}\\ &= 2\sin\frac{\alpha+\beta}{2}\cos\frac{\alpha-\beta}{2},\end{aligned}$$

which is one of the **sum-to-product formulas**. Similarly, we obtain various forms of the sum-to-product formulas and **difference-to-product formulas**.

Example 1.4. Let a and b be nonnegative real numbers.

(a) Prove that there is a real number x such that $\sin x + a\cos x = b$ if and only if $a^2 - b^2 + 1 \geq 0$.

(b) If $\sin x + a\cos x = b$, express $|a\sin x - \cos x|$ in terms of a and b.

Solution: To establish (a), we prove a more general result.

(a) Let m, n, and ℓ be real numbers such that $m^2 + n^2 \neq 0$. We will prove that there is a real number x such that

$$m\sin x + n\cos x = \ell \qquad (*)$$

if and only if $m^2 + n^2 \geq \ell^2$.

Indeed, we can rewrite equation $(*)$ in the following form:

$$\frac{m}{\sqrt{m^2+n^2}}\sin x + \frac{n}{\sqrt{m^2+n^2}}\cos x = \frac{\ell}{\sqrt{m^2+n^2}}.$$

Point $\left(\frac{m}{\sqrt{m^2+n^2}}, \frac{n}{\sqrt{m^2+n^2}}\right)$ lies on the unit circle. There is a unique real number α with $0 \leq \alpha < 2\pi$ such that

$$\cos\alpha = \frac{m}{\sqrt{m^2+n^2}} \quad\text{and}\quad \sin\alpha = \frac{n}{\sqrt{m^2+n^2}}.$$

The **addition and subtraction formulas** yield

$$\sin(x+\alpha) = \cos\alpha\sin x + \sin\alpha\cos x = \frac{\ell}{\sqrt{m^2+n^2}},$$

which is solvable in x if and only if $-1 \leq \frac{\ell}{\sqrt{m^2+n^2}} \leq 1$, that is, if and only if $\ell^2 \leq m^2 + n^2$. Setting $m = a$, $n = 1$, and $\ell = c$ gives the desired result.

(b) By the relations

$$\begin{aligned}
a^2+1 &= (\sin^2 x + \cos^2 x)(a^2+1) \\
&= (\sin^2 x + 2a\sin x\cos x + a^2\cos^2 x) \\
&\quad + (a^2\sin^2 x - 2a\sin x\cos x + \cos^2 x) \\
&= (\sin x + a\cos x)^2 + (a\sin x - \cos x)^2,
\end{aligned}$$

we conclude that $|a\sin x - \cos x| = \sqrt{a^2 - b^2 + 1}$. ■

Graphs of Trigonometric Functions

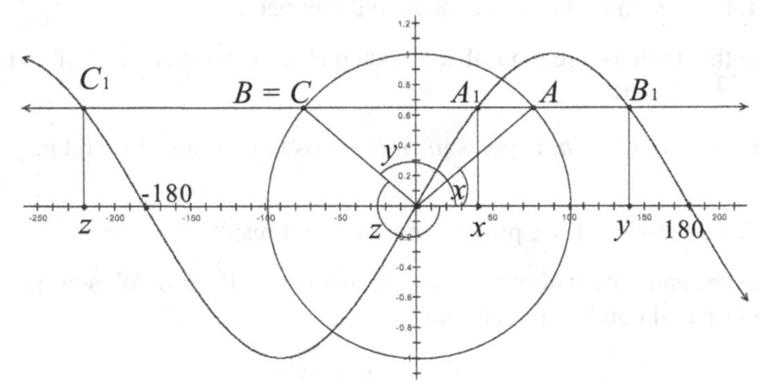

Figure 1.12.

We set the units of the x axis to be degrees. The graph of $y = \sin x$ looks like a wave, as shown in Figure 1.12. (This is only part of the graph. The graph extends infinitely in both directions along the x axis.) For example, the point $A = (1, x^\circ)$ corresponds to the point $A_1 = (x, \sin x)$ on the curve $y = \sin x$. If two points B_1 and C_1 are 360 from each other in the x direction, then they have the same y value, and they correspond to the same point $B = C$ on the unit circle. (This is the correspondence of the identity $\sin(x^\circ + 360^\circ) = \sin x^\circ$.) Also, the graph is symmetric about line $x = 90$. (This corresponds to the identity $\sin(90^\circ - x^\circ) = \sin(90^\circ + x^\circ)$.) The identity $\sin(-x^\circ) = -\sin x^\circ$ indicates that the graph $y = \sin x$ is symmetric with respect to the origin; that is, the sine is an **odd function**.

A function $y = f(x)$ is **sinusoidal** if it can be written in the form $y = f(x) = a\sin[b(x+c)] + d$ for real constants a, b, c, and d. In particular, because $\cos x^\circ = \sin(x^\circ + 90^\circ)$, $y = \cos x$ is sinusoidal (Figure 1.13). For any integer k, the graph of $y = \cos x$ is a $(90 + 360k)$-unit shift to the left (or a $(270 + 360k)$-unit shift to the right) of the graph of $y = \sin x$. Because $\cos x^\circ = \cos(-x^\circ)$, the cosine is an **even function**, and so its graph is symmetric about the y axis.

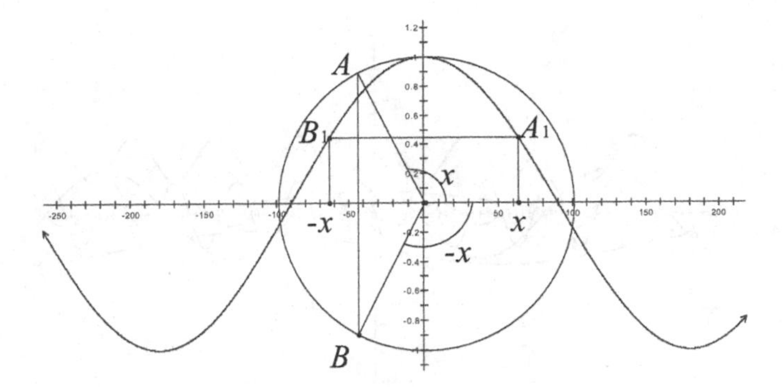

Example 1.5. Let f be an odd function defined on the real numbers such that for $x \geq 0$, $f(x) = 3\sin x + 4\cos x$. Find $f(x)$ for $x < 0$. (See Figure 1.14.)

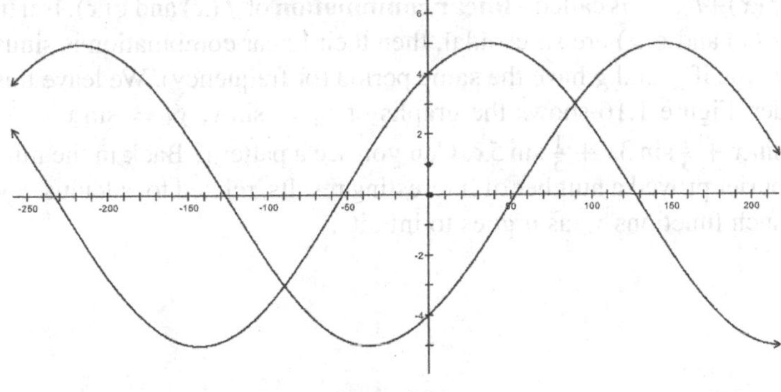

Figure 1.14.

Solution: Because f is odd, $f(x) = -f(-x)$. For $x < 0$, $-x > 0$ and $f(-x) = 3\sin(-x) + 4\cos(-x) = -3\sin x + 4\cos x$ by definition. Hence, for $x < 0$, $f(x) = -(-3\sin x + 4\cos x) = 3\sin x - 4\cos x$. (It seems that $y = 3\sin x + 4\cos x$ might be sinusoidal; can you prove or disprove this?) ■

For a sinusoidal function $y = a\sin[b(x + c)] + d$, it is important to note the roles played by the constants a, b, c and d in its graph. Generally speaking, a is the **amplitude** of the curve, b is related to the period of the curve, c is related to the horizontal shift of the curve, and d is related to the vertical shift of the curve. To get a clearer picture, the reader might want to match the functions $y = \sin 3x$, $y = 2\cos\frac{x}{3}$, $y = 3\sin 4x$, $y = 4\cos(x - 30°)$, $y = \frac{3}{2}\sin\frac{x}{2} - 3$, and $y = 2\sin[3(x + 40°)] + 5$ with the curves shown in Figure 1.15.

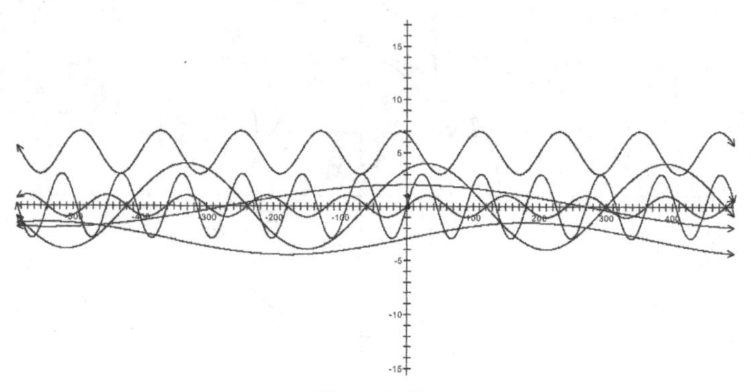

Figure 1.15.

We leave it to the reader to show that if a, b, c, and d are real constants, then the functions $y = a\cos(bx+c)+d$, $y = a\sin x+b\cos x$, $y = a\sin^2 x$, and $y = b\cos^2 x$ are sinusoidal. Let $f(x)$ and $g(x)$ be two functions. For real constants a and b, the function $af(x)+bg(x)$ is called a **linear combination** of $f(x)$ and $g(x)$. Is it true that if both of $f(x)$ and $g(x)$ are sinusoidal, then their linear combination is sinusoidal? In fact, it is true if f and g have the same period (or frequency). We leave this proof to the reader. Figure 1.16 shows the graphs of $y_1 = \sin x$, $y_3 = \sin x + \frac{1}{3}\sin 3x$, and $y_5 = \sin x + \frac{1}{3}\sin 3x + \frac{1}{5}\sin 5x$. Can you see a pattern? Back in the nineteenth century, Fourier proved a number of interesting results, related to calculus, about the graphs of such functions y_n as n goes to infinity.

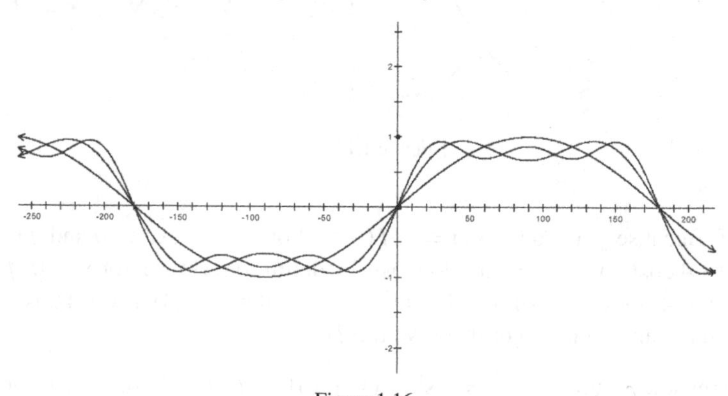

Figure 1.16.

The graphs of $y = \tan x$ and $y = \cot x$ are not continuous, because $\tan x$ is not defined for $x = (2k+1)\cdot 90°$ and $\cot x$ is not defined for $x = k \cdot 180°$, where k is an integer. The graph of $\tan x$ has vertical **asymptotes** at $x = (2k+1)\cdot 90°$; that is, as x approaches $k \cdot 180°$, the values of $\tan x$ grow large in absolute value, and so the graph of the tangent function moves closer and closer to the asymptote, as shown in Figure 1.17. Similarly, the graph of $\cot x$ has vertical asymptotes at $x = k \cdot 180°$.

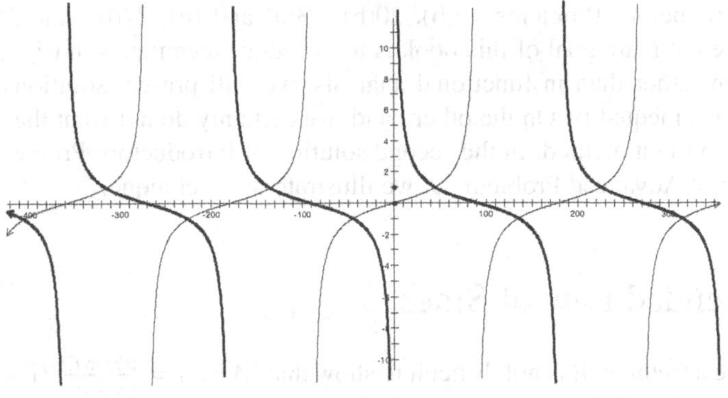

Figure 1.17.

A function $f(x)$ is **concave up (down)** on an interval $[a, b]$ if the graph of $f(x)$ lies under (over) the line connecting $(a_1, f(a_1))$ and $(b_1, f(b_1))$ for all

$$a \leq a_1 < x < b_1 \leq b.$$

Functions that are concave up and down are also called **convex** and **concave**, respectively. In other words, the graph of a concave up function looks like a bowl that *holds water*, while the graph of a concave down function looks like a bowl that *spills water*.

If f is concave up on an interval $[a, b]$ and $\lambda_1, \lambda_2, \ldots, \lambda_n$ (λ – Greek "lambda") are nonnegative numbers with sum equal to 1, then

$$\lambda_1 f(x_1) + \lambda_2 f(x_2) + \cdots + \lambda_n f(x_n) \geq f(\lambda_1 x_1 + \lambda_2 x_2 + \cdots + \lambda_n x_n)$$

for any $x_1, x_2, \ldots, x_n$ in the interval $[a, b]$. If the function is concave down, the inequality is reversed. This is **Jensen's inequality**. Jensen's inequality says that the output of a convex function at the weighted average of a group of inputs is less than or equal to the same weighted average of the outputs of the function at the group of inputs.

It is not difficult to see that $y = \sin x$ is concave down for $0° \leq x \leq 180°$ and $y = \tan x$ is concave up for $0° \leq x < 90°$. By Jensen's inequality, for triangle ABC, we have

$$\frac{1}{3}\sin A + \frac{1}{3}\sin B + \frac{1}{3}\sin C \leq \sin\frac{A+B+C}{3} = \frac{\sqrt{3}}{2},$$

or $\sin A + \sin B + \sin C \leq \frac{3\sqrt{3}}{2}$, which is Introductory Problem 28(c). Similarly, we have $\tan A + \tan B + \tan C \geq 3\sqrt{3}$. For those who know calculus, convexity of a function is closely related to the second derivative of the function. We can also use the natural logarithm function to change products into sums, and then apply Jensen's inequality. This technique will certainly be helpful in solving problems such as Introductory Problems 19(b), 20(b), 23(a) and (d), 27(b), and 28(b) and (c). Because the main goal of this book is to introduce techniques in trigonometric computation rather than in functional analysis, we will present solutions without using Jensen's inequality. On the other hand, we certainly do not want the reader to miss this important method. In the second solution of Introductory Problem 51 and the solution of Advanced Problem 39, we illustrate this technique.

The Extended Law of Sines

Let ABC be a triangle. It is not difficult to show that $[ABC] = \frac{ab\sin C}{2}$. (For a proof, see the next section.) By symmetry, we have

$$[ABC] = \frac{ab\sin C}{2} = \frac{bc\sin A}{2} = \frac{ca\sin B}{2}.$$

Dividing all sides of the last equation by $\frac{abc}{2}$ gives the **law of sines**:

$$\frac{\sin A}{a} = \frac{\sin B}{b} = \frac{\sin C}{c} \quad \text{or} \quad \frac{a}{\sin A} = \frac{b}{\sin B} = \frac{c}{\sin C}.$$

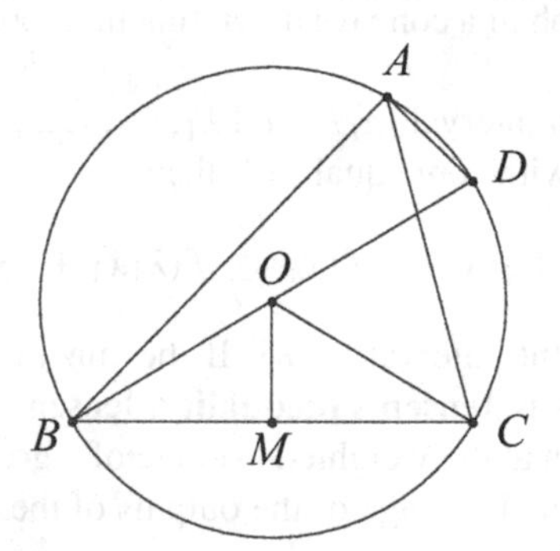

Figure 1.18.

The common ratio $\frac{a}{\sin A}$ has a significant geometric meaning. Let ω be the circumcircle of triangle ABC, and let O and R be the center and radius of ω, respectively. (See Figure 1.18.) Then $\angle BOC = 2\angle CAB$. Let M be the midpoint of segment BC. Because triangle OBC is isosceles with $|OB| = |OC| = R$, it follows that $OM \perp BC$ and $\angle BOM = \angle COM = \angle CAB$. In the right triangle BMO, $|BM| = |OB|\sin A$; that is, $\frac{a}{\sin A} = \frac{2|BM|}{\sin A} = 2|OB| = 2R$. Hence, we obtain the **extended law of sines**: In a triangle ABC with circumradius equal to R,

$$\frac{a}{\sin A} = \frac{b}{\sin B} = \frac{c}{\sin A} = 2R.$$

Note that this fact can also be obtained by extending ray OB to meet ω at D, and then working on right triangle ABD.

A direct application of the law of sines is to prove the **angle-bisector theorem**: Let ABC be a triangle (Figure 1.19), and let D be a point on segment BC such that $\angle BAD = \angle CAD$. Then

$$\frac{|AB|}{|AC|} = \frac{|BD|}{|CD|}.$$

Applying the law of sines to triangle ABD gives

$$\frac{|AB|}{\sin \angle ADB} = \frac{|BD|}{\sin \angle BAD}, \quad \text{or} \quad \frac{|AB|}{|BD|} = \frac{\sin \angle ADB}{\sin \angle BAD}.$$

Similarly, applying the law of sines to triangle ACD gives $\frac{|AC|}{|CD|} = \frac{\sin \angle ADC}{\sin \angle CAD}$. Because $\sin \angle ADB = \sin \angle ADC$ and $\sin \angle BAD = \sin \angle CAD$, it follows that $\frac{|AB|}{|BD|} = \frac{|AC|}{|CD|}$, as desired.

This theorem can be extended to the situation in which AD_1 is the external bisector of the triangle (see Figure 1.19). We leave it to the reader to state and prove this version of the theorem.

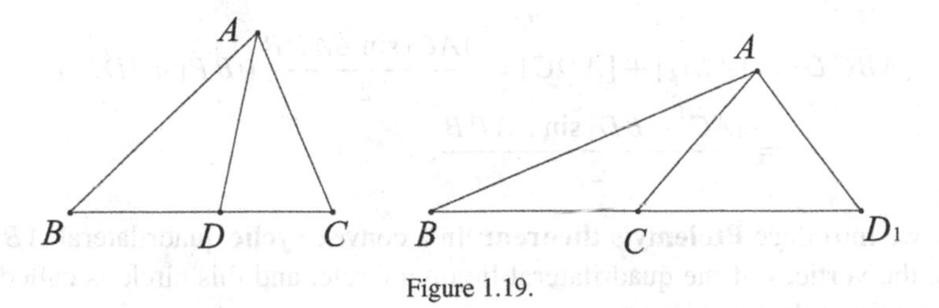

Figure 1.19.

Area and Ptolemy's Theorem

Let ABC be a triangle, and let D be the foot of the perpendicular line segment from A to line BC (Figure 1.20). Then $[ABC] = \frac{|BC|\cdot|AD|}{2}$. Note that $|AD| = |AB| \sin B$. Thus $[ABC] = \frac{|BC|\cdot|AB| \sin B}{2} = \frac{ac \sin B}{2}$.

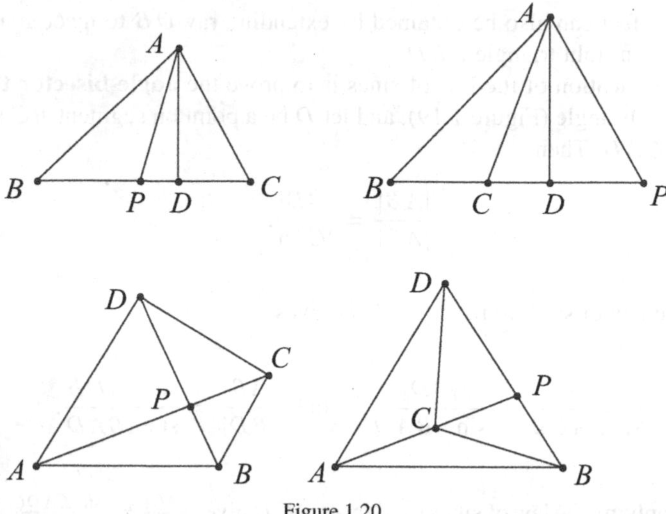

Figure 1.20.

In general, if P is a point on segment BC, then $|AD| = |AP| \sin \angle APB$. Hence $[ABC] = \frac{|AP|\cdot|BC| \sin \angle APB}{2}$. More generally, let $ABCD$ be a quadrilateral (not necessarily convex), and let P be the intersection of diagonals AC and BD, as shown in Figure 1.20. Then $[ABC] = \frac{|AC|\cdot|BP| \sin \angle APB}{2}$ and $[ADC] = \frac{|AC|\cdot|DP| \sin \angle APD}{2}$. Because $\angle APB + \angle APD = 180°$, it follows that $\sin \angle APB = \sin \angle APD$ and

$$\begin{aligned}[ABCD] = [ABC] + [ADC] &= \frac{|AC| \sin \angle APB}{2}(|BP| + |DP|) \\ &= \frac{|AC| \cdot |BD| \sin \angle APB}{2}.\end{aligned}$$

Now we introduce **Ptolemy's theorem**: In a convex **cyclic** quadrilateral $ABCD$ (that is, the vertices of the quadrilateral lie on a circle, and this circle is called the **circumcircle** of the quadrilateral),

$$|AC| \cdot |BD| = |AB| \cdot |CD| + |AD| \cdot |BC|.$$

There are many proofs of this very important theorem. Our proof uses areas. The product $|AC| \cdot |BD|$ is closely related to $[ABCD]$. Indeed,

$$[ABCD] = \frac{1}{2} \cdot |AC| \cdot |BD| \sin \angle APB,$$

where P is the intersection of diagonals AC and BD. (See Figure 1.21.)

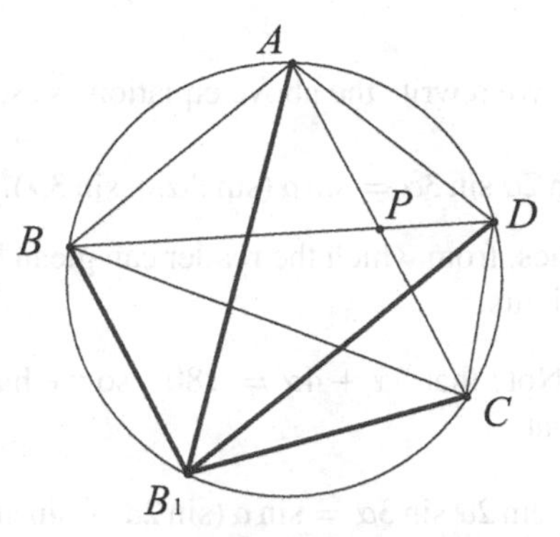

Figure 1.21.

Hence, we want to express the products $|AB| \cdot |CD|$ and $|BC| \cdot |DA|$ in terms of areas. To do so, we reflect B across the perpendicular bisector of diagonal AC. Let B_1 be the image of B under the reflection. Then ABB_1C is an isosceles trapezoid with $BB_1 \parallel AC$, $|AB| = |CB_1|$, and $|AB_1| = |CB|$. Also note that B_1 lies on the circumcircle of $ABCD$. Furthermore, $\widehat{AB} = \widehat{CB_1}$, and so

$$\angle B_1AD = \frac{\widehat{B_1D}}{2} = \frac{\widehat{B_1C} + \widehat{CD}}{2} = \frac{\widehat{AB} + \widehat{CD}}{2} = \angle APB.$$

Because AB_1CD is cyclic, $\angle B_1AD + \angle B_1CD = 180°$. Thus $\sin \angle B_1AD = \sin \angle B_1CD = \sin \angle APB$. Because of symmetry, we have

$$\begin{aligned}
[ABCD] &= [ABC] + [ACD] = [AB_1C] + [ACD] \\
&= [AB_1CD] = [AB_1D] + [CB_1D] \\
&= \frac{1}{2} \cdot |AB_1| \cdot |AD| \sin \angle B_1AD + \frac{1}{2} \cdot |CB_1| \cdot |CD| \sin \angle B_1CD \\
&= \frac{1}{2} \cdot \sin \angle APB(|BC| \cdot |AD| + |AB| \cdot |CD|).
\end{aligned}$$

By calculating $[ABCD]$ in two different ways, we establish

$$\frac{1}{2} \cdot |AC| \cdot |BD| \sin \angle APB = \frac{1}{2} \cdot \sin \angle APB(|BC| \cdot |AD| + |AB| \cdot |CD|),$$

or $|AC| \cdot |BD| = |BC| \cdot |AD| + |AB| \cdot |CD|$, completing the proof of the theorem.

In Introductory Problem 52, we discuss many interesting properties of the special angle $\frac{180°}{7}$. The following is the first of these properties.

Example 1.6. Prove that

$$\csc \frac{180°}{7} = \csc \frac{360°}{7} + \csc \frac{540°}{7}.$$

Solution: Let $\alpha = \frac{180^\circ}{7}$. We rewrite the above equation as $\csc\alpha = \csc 2\alpha + \csc 3\alpha$, or

$$\sin 2\alpha \sin 3\alpha = \sin\alpha(\sin 2\alpha + \sin 3\alpha).$$

We present two approaches, from which the reader can glean both algebraic computation and geometric insights.

- **First Approach:** Note that $3\alpha + 4\alpha = 180^\circ$, so we have $\sin 3\alpha = \sin 4\alpha$. It suffices to show that

  $$\sin 2\alpha \sin 3\alpha = \sin\alpha(\sin 2\alpha + \sin 4\alpha).$$

  By the **addition and subtraction formulas**, we have $\sin 2\alpha + \sin 4\alpha = 2\sin 3\alpha \cos\alpha$. Then the desired result reduces to $\sin 2\alpha = 2\sin\alpha\cos\alpha$, which is the **double-angle formula** for the sine function.

- **Second Approach:** Consider a regular heptagon $A_1A_2\ldots A_7$ inscribed in a circle of radius $R = \frac{1}{2}$ (Figure 1.22). Then each arc $\widehat{A_iA_{i+1}}$ has measure $\frac{360^\circ}{7} = 2\alpha$.

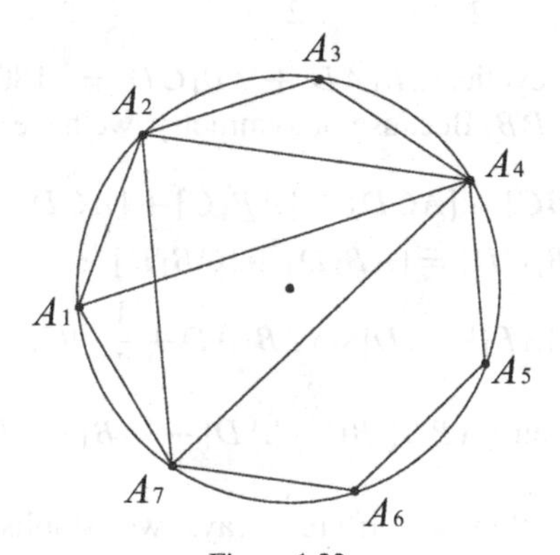

Figure 1.22.

By the **extended law of sines**, we have $|A_1A_2| = |A_1A_7| = 2R\sin\alpha = \sin\alpha$, $|A_2A_4| = |A_2A_7| = \sin 2\alpha$, and $|A_1A_4| = |A_4A_7| = \sin 3\alpha$. Applying Ptolemy's theorem to the cyclic quadrilateral $A_1A_2A_4A_7$ gives

$$|A_1A_4| \cdot |A_2A_7| = |A_1A_2| \cdot |A_4A_7| + |A_2A_4| \cdot |A_7A_1|;$$

that is,

$$\sin 2\alpha \sin 3\alpha = \sin\alpha(\sin 2\alpha + \sin 3\alpha).$$

Existence, Uniqueness, and Trigonometric Substitutions

The fact that $\sin\alpha = \sin\beta$, for $\alpha + \beta = 180^\circ$, has already helped us in many places. It also helped us to explain why either side-side-angle (SSA) or area-side-side information is not enough to determine the unique structure of a triangle.

Example 1.7. Let ABC be a triangle.

(a) Suppose that $[ABC] = 10\sqrt{3}$, $|AB| = 8$, and $|AC| = 5$. Find all possible values of $\angle A$.

(b) Suppose that $|AB| = 5\sqrt{2}$, $|BC| = 5\sqrt{3}$, and $\angle C = 45^\circ$. Find all possible values of $\angle A$.

(c) Suppose that $|AB| = 5\sqrt{2}$, $|BC| = 5$, and $\angle C = 45^\circ$. Find all possible values of $\angle A$.

(d) Suppose that $|AB| = 5\sqrt{2}$, $|BC| = 10$, and $\angle C = 45^\circ$. Find all possible values of $\angle A$.

(e) Suppose that $|AB| = 5\sqrt{2}$, $|BC| = 15$, and $\angle C = 45^\circ$. Find all possible values of $\angle A$.

Solution:

(a) Note that $b = |AC| = 5$, $c = |AB| = 8$, and $[ABC] = \frac{1}{2}bc\sin A$. Thus $\sin A = \frac{\sqrt{3}}{2}$, and $A = 60^\circ$ or 120° (A_1 and A_2 in Figure 1.23).

(b) By the **law of sines**, we have $\frac{|BC|}{\sin A} = \frac{|AB|}{\sin C}$, or $\sin A = \frac{\sqrt{3}}{2}$. Hence $A = 60^\circ$ or 120°.

(c) By the law of sines, we have $\frac{|BC|}{\sin A} = \frac{|AB|}{\sin C}$, or $\sin A = \frac{1}{2}$. Hence $A = 30^\circ$ only (A_3 in Figure 1.23)! (Why?)

(d) By the law of sines, we have $\sin A = 1$, and so $A = 90^\circ$. (A_4 in Figure 1.23)

(e) By the law of sines, we have $\sin A = \frac{3}{2}$, which is impossible. We conclude that there is no triangle satisfying the conditions of the problem. ∎

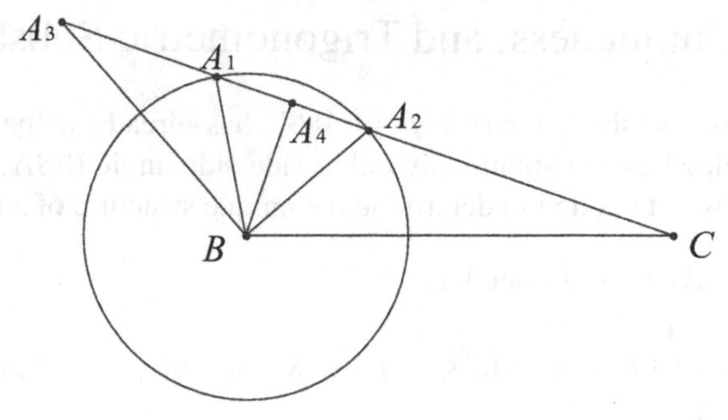

Figure 1.23.

Example 1.8. [AMC12 2001] In triangle ABC, $\angle ABC = 45°$. Point D is on segment BC such that $2|BD| = |CD|$ and $\angle DAB = 15°$. Find $\angle ACB$.

First Solution: We construct this triangle in the following way: Fix segment BC, choose point D on segment BC such that $2|BD| = |CD|$ (Figure 1.24, left), and construct ray BP such that $\angle PBC = 45°$. Let A be a point on ray BP that moves from B in the direction of the ray. It is not difficult to see that $\angle DAB$ decreases as A moves away from B. Hence, there is a unique position for A such that $\angle DAB = 15°$. This completes our construction of triangle ABC.

This figure brings to mind the proof of the **angle-bisector theorem**. We apply the **law of sines** to triangles ACD and ABC. Set $\alpha = \angle CAD$. Note that $\angle CDA = \angle CBA + \angle DAB = 60°$. We have

$$\frac{|CD|}{\sin\alpha} = \frac{|CA|}{\sin 60°} \quad \text{and} \quad \frac{|BC|}{\sin(\alpha + 15°)} = \frac{|CA|}{\sin 45°}.$$

Dividing the first equation by the second equations gives

$$\frac{|CD|\sin(\alpha + 15°)}{|BC|\sin\alpha} = \frac{\sin 45°}{\sin 60°}.$$

Note that $\frac{|CD|}{|BC|} = \frac{2}{3} = \left(\frac{\sin 45°}{\sin 60°}\right)^2$. It follows that

$$\left(\frac{\sin 45°}{\sin 60°}\right)^2 = \frac{\sin\alpha}{\sin(\alpha + 15°)} \cdot \frac{\sin 45°}{\sin 60°}.$$

It is clear that $\alpha = 45°$ is a solution of the above equation. By the uniqueness of our construction, it follows that $\angle ABC = 45°$, $\angle CAB = 60°$, and $\angle ACB = 75°$.

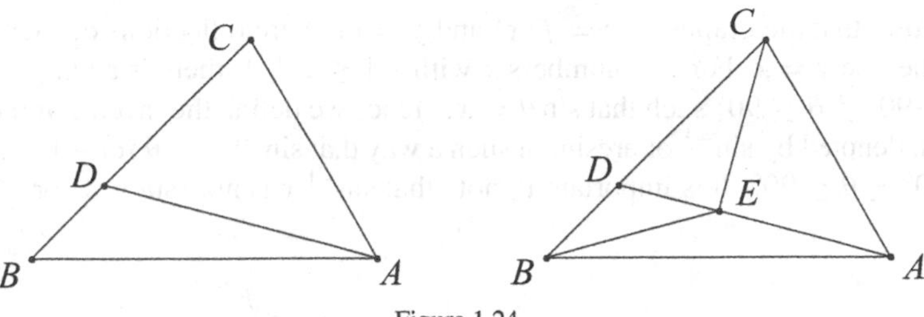

Figure 1.24.

Second Solution: Note that $\angle CDA = 60°$ and that $\sin 30° = \frac{1}{2}$. We construct point E on segment AD (Figure 1.24, right) such that $CE \perp AD$. Then in triangle CDE, $\angle DCE = 30°$ and $|DE| = |CD| \sin \angle DCE$, or $|CD| = 2|DE|$. Thus triangle BDE is isosceles with $|DE| = |DB|$, implying that $\angle DBE = \angle DEB = 30°$. Consequently, $\angle CBE = \angle BCE = 30°$ and $\angle EBA = \angle EAB = 15°$, and so triangles BCE and BAE are both isosceles with $|CE| = |BE| = |EA|$. Hence the right triangle AEC is isosceles; that is, $\angle ACE = \angle EAC = 45°$. Therefore, $\angle ACB = \angle ACE + \angle ECB = 75°$. ■

For a function $f : A \to B$, if $f(A) = B$, then f is said to be **surjective** (or **onto**); that is, every $b \in B$ is the image under f of some $a \in A$. If every two distinct elements a_1 and a_2 in A have distinct images, then f is **injective** (or **one-to-one**). If f is both injective and surjective, then f is **bijective** (or a **bijection** or a **one-to-one correspondence**).

The sine and cosine functions are functions from the set of angles to the real numbers. The images of the two functions are the real numbers between -1 and 1. For a point $P = (x, y)$ with polar coordinates $(1, \theta)$ on the unit circle, it is clear that the values $x = \cos\theta$ and $y = \sin\theta$ vary continuously from -1 to 1, taking on all intermediate values. Hence the functions are surjective functions from the set of angles to the interval $[-1, 1]$. On the other hand, these two functions are not one-to-one. It is not difficult to see that the sine function is a bijection between the set of angles α with $-90° \le \alpha \le 90°$ and the interval $[-1, 1]$, and that the cosine function is a bijection between the set of angles α with $0° \le \alpha \le 180°$ and the interval $[-1, 1]$. For abbreviation, we can write that $\sin : [-90°, 90°] \to [-1, 1]$ is a bijection. It is also not difficult to see that the tangent function is a bijection between the set of angles α with $-90° < \alpha < 90°$ ($0° < \alpha < 90°$, or $0° \le \alpha < 90°$) and the set of real numbers (positive real numbers, or nonnegative real numbers).

Two functions f and g are **inverses** of each other if $f(g(x)) = x$ for all x in the domain of g and $g(f(x)) = x$ for all x in the domain of f. If the function f is one-to-one and onto, then it is not difficult to see that f has an inverse. For a pair of functions f and g that are inverses of each other, if $y = f(x)$, then $g(y) = g(f(x)) = x$; that is, if (a, b) lies on the graph of $y = f(x)$, then (b, a) lies on the graph of $y = g(x)$.

It follows that the graphs of $y = f(x)$ and $y = g(x)$ are reflections of each other across the line $y = x$. For real numbers x with $-1 \le x \le 1$, there is a unique angle θ with $-90^\circ \le \theta \le 90^\circ$ such that $\sin\theta = x$. Hence we define the inverse of the sine function, denoted by $\sin^{-1}$ or arcsin, in such a way that $\sin^{-1} x = \theta$ for $-1 \le x \le 1$ and $-90^\circ \le \theta \le 90^\circ$. It is important to note that $\sin^{-1} x$ is not $(\sin x)^{-1}$ or $\frac{1}{\sin x}$.

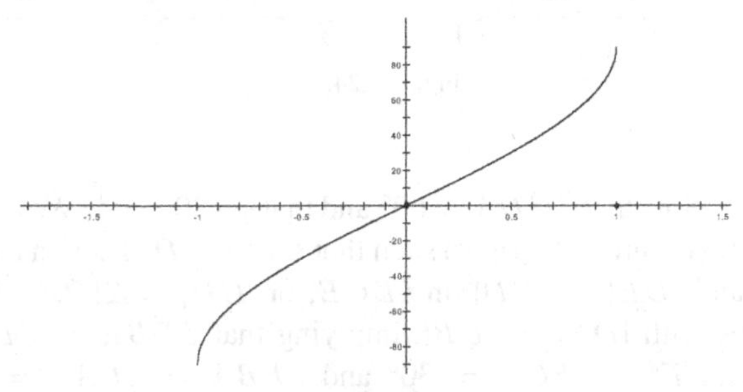

Figure 1.25.

Similarly, we can define the inverse functions of $\tan x$ and $\cot x$. They are denoted by $\tan^{-1} x$ (or $\arctan x$) and $\cot^{-1} x$. Both functions have domain $\mathbb{R}$. Their ranges are $\{\theta \mid -90^\circ < \theta < 90^\circ\}$ and $\{\theta \mid -90^\circ < \theta \le 90^\circ, \theta \ne 0^\circ\}$. They are both one-to-one functions and onto functions. Their graphs are shown below. Note that $y = \arctan x$ has two horizontal asymptotes $y = 90$ and $y = -90$. Note also that $y = \cot^{-1} x$ has two pieces, and both of them are asymptotic to the line $y = 0$. See Figure 1.26.

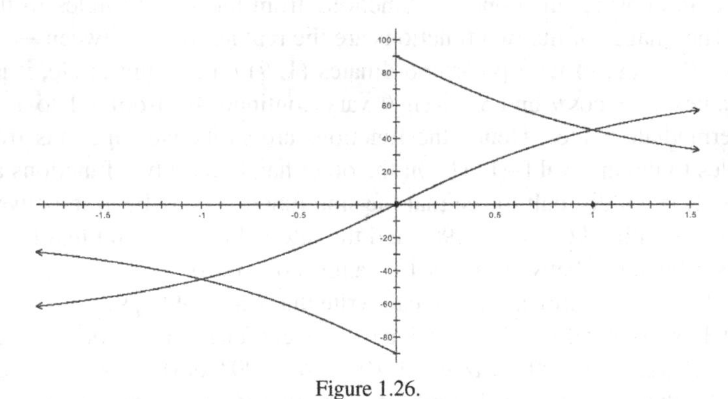

Figure 1.26.

Note that $y = \cos x$ is one-to-one and onto from $\{\theta \mid 0^\circ \le \theta \le 180^\circ\}$ to the interval $[-1, 1]$. While the domain of $\cos^{-1} x$ (or $\arccos x$) is the same as that of $\arcsin x$, the range of $\cos^{-1} x$ is $\{\theta \mid 0^\circ \le \theta \le 180^\circ\}$. See Figure 1.27.

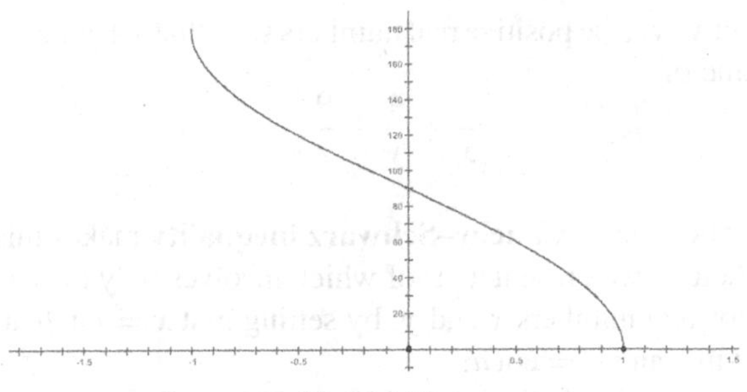

Figure 1.27.

The graphs of $y = \sin^{-1}(\sin x)$, $y = \cos^{-1}(\cos x)$, $y = \sin(\sin^{-1} x)$, $y = \cos(\cos^{-1} x)$, $y = \cos^{-1}(\sin x)$, $y = \sin^{-1}(\cos x)$, $y = \sin(\cos^{-1} x)$, and $y = \cos(\sin^{-1} x)$ are interesting and important. We leave it to the reader to complete these graphs. We close this section with three examples of trigonometric substitution.

Example 1.9. Let $x_0 = 2003$, and let $x_{n+1} = \frac{1+x_n}{1-x_n}$ for $n \geq 1$. Compute x_{2004}.

Solution: With a little algebraic computation, we can show that this sequence has a period of 4; that is, $x_{n+4} = x_n$ for all $n \geq 1$. But why? We reveal the secret with trigonometric substitution; that is, we define α_n with $-90° < \alpha_n < 90°$ such that $\tan \alpha_n = x_n$. It is clear that if x_n is a real number, such an α_n is unique, because $\tan : (-90°, 90°) \to \mathbb{R}$ is a bijection. Because $1 = \tan 45°$, we can rewrite the given condition as

$$\tan \alpha_{n+1} = \frac{\tan 45° + \tan \alpha_n}{1 - \tan 45° \tan \alpha_n} = \tan(45° + \alpha_n),$$

by the **addition and subtraction formulas**. Consequently, $\alpha_{n+1} = 45° + \alpha_n$, or $\alpha_{n+1} = 45° + \alpha_n - 180°$ (because tan has a period of 180°). In any case, it is not difficult to see that $\alpha_{n+4} = \alpha_n + k \cdot 180°$ for some integer k. Therefore, $x_{n+4} = \tan \alpha_{n+4} = \tan \alpha_n = x_n$; that is, the sequence $\{x_n\}_{n\geq 0}$ has period 4, implying that $x_{2004} = x_0 = 2003$. ■

Example 1.10. Prove that among any five distinct real numbers there are two, a and b, such that $|ab + 1| > |a - b|$.

Solution: Write the numbers as $\tan x_k$, where $-90° < x_k < 90°$, $k = 1, 2, 3, 4, 5$. We consider the intervals $(-90°, -45°]$, $(-45°, 0°]$, $(0°, 45°]$, and $(45°, 90°]$. By the **pigeonhole principle**, at least two of x_1, x_2, x_3, x_4, x_5 lie in the same interval, say x_i and x_j. Then $|x_i - x_j| < 45°$, and setting $a = \tan x_i$ and $b = \tan x_j$, we get

$$\left|\frac{a-b}{1+ab}\right| = \left|\frac{\tan x_i - \tan x_j}{1 + \tan x_i \tan x_j}\right| = |\tan(x_i - x_j)| < \tan 45° = 1,$$

and hence the conclusion follows. ■

Example 1.11. Let x, y, z be positive real numbers such that $x+y+z=1$. Determine the minimum value of

$$\frac{1}{x}+\frac{4}{y}+\frac{9}{z}.$$

Solution: An application of **Cauchy–Schwarz inequality** makes this is a one-step problem. Nevertheless, we present a proof which involves only the easier inequality $x^2+y^2 \geq 2xy$ for real numbers x and y, by setting first $x=\tan b$ and $y=2\tan b$ and second $x=\tan a$ and $y=\cot a$.

Clearly, z is a real number in the interval $[0, 1]$. Hence there is an angle a such that $z=\sin^2 a$. Then $x+y=1-\sin^2 a=\cos^2 a$, or $\frac{x}{\cos^2 a}+\frac{y}{\cos^2 a}=1$. For an angle b, we have $\cos^2 b+\sin^2 b=1$. Hence, we can set $x=\cos^2 a\cos^2 b$, $y=\cos^2 a\sin^2 b$ for some angle b. It suffices to find the minimum value of

$$P=\sec^2 a\sec^2 b+4\sec^2 a\csc^2 b+9\csc^2 a,$$

or

$$P=(\tan^2 a+1)(\tan^2 b+1)+4(\tan^2 a+1)(\cot^2 b+1)+9(\cot^2 a+1).$$

Expanding the right-hand side gives

$$\begin{aligned} P &= 14+5\tan^2 a+9\cot^2 a+(\tan^2 b+4\cot^2 b)(1+\tan^2 a) \\ &\geq 14+5\tan^2 a+9\cot^2 a+2\tan b\cdot 2\cot b\left(1+\tan^2 a\right) \\ &= 18+9(\tan^2 a+\cot^2 a)\geq 18+9\cdot 2\tan a\cot a=36. \end{aligned}$$

Equality holds when $\tan a=\cot a$ and $\tan b=2\cot b$, which implies that $\cos^2 a=\sin^2 a$ and $2\cos^2 b=\sin^2 b$. Because $\sin^2\theta+\cos^2\theta=1$, equality holds when $\cos^2 a=\frac{1}{2}$ and $\cos^2 b=\frac{1}{3}$; that is, $x=\frac{1}{6}$, $y=\frac{1}{3}$, $z=\frac{1}{2}$. ■

Ceva's Theorem

A **cevian** of a triangle is any segment joining a vertex to a point on the opposite side.

[**Ceva's Theorem**] Let AD, BE, CF be three cevians of triangle ABC. The following are equivalent (see Figure 1.28):

(1) AD, BE, CF are concurrent; that is, these lines pass a common point;

(2) $$\frac{\sin\angle ABE}{\sin\angle DAB}\cdot\frac{\sin\angle BCF}{\sin\angle EBC}\cdot\frac{\sin\angle CAD}{\sin\angle FCA}=1;$$

(3) $\dfrac{|AF|}{|FB|} \cdot \dfrac{|BD|}{|DC|} \cdot \dfrac{|CE|}{|EA|} = 1.$

We will show that (1) implies (2), (2) implies (3), and then (3) implies (1).

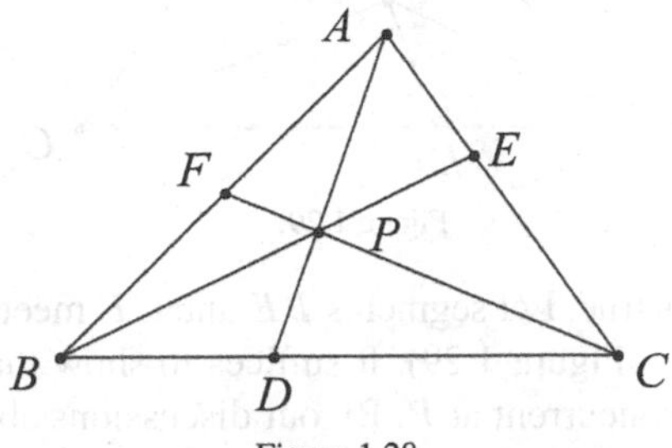

Figure 1.28.

Assume that part (1) is true. We assume that segments AD, BE, and CF meet at point P. Applying the **law of sines** to triangle ABP yields

$$\frac{\sin \angle ABE}{\sin \angle DAB} = \frac{\sin \angle ABP}{\sin \angle PAB} = \frac{|AP|}{|BP|}.$$

Similarly, applying law of sines to triangle BCP and CAP gives

$$\frac{\sin \angle BCF}{\sin \angle EBC} = \frac{|BP|}{|CP|} \quad \text{and} \quad \frac{\sin \angle CAD}{\sin \angle FCA} = \frac{|CP|}{|AP|}.$$

Multiplying the last three identities gives part (2).

Assume that part (2) is true. Applying the law of sines to triangles ABD and ACD gives

$$\frac{|AB|}{|BD|} = \frac{\sin \angle ADB}{\sin \angle DAB} \quad \text{and} \quad \frac{|DC|}{|CA|} = \frac{\sin \angle CAD}{\sin \angle ADC}.$$

Because $\angle ADC + \angle ADB = 180°$, we have $\sin \angle ADB = \sin \angle ADC$. Multiplying the above identities gives

$$\frac{|DC|}{|BD|} \cdot \frac{|AB|}{|CA|} = \frac{\sin \angle CAD}{\sin \angle DAB}.$$

Likewise, we have

$$\frac{|AE|}{|EC|} \cdot \frac{|BC|}{|AB|} = \frac{\sin \angle ABE}{\sin \angle EBC} \quad \text{and} \quad \frac{|BF|}{|FA|} \cdot \frac{|CA|}{|BC|} = \frac{\sin \angle BCF}{\sin \angle FCA}.$$

Multiplying the last three identities gives part (3).

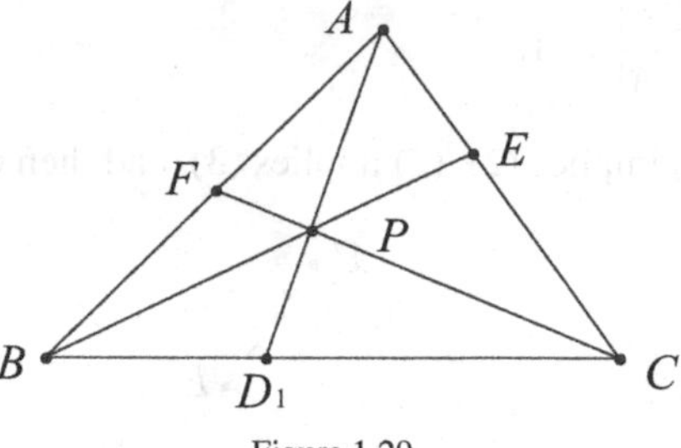

Figure 1.29.

Assume that part (3) is true. Let segments BE and CF meet at P, and let ray AP meet segment BC at D_1 (Figure 1.29). It suffices to show that $D = D_1$. Cevians AD_1, BE, and CF are concurrent at P. By our discussions above, we have

$$\frac{|AF|}{|FB|} \cdot \frac{|BD_1|}{|D_1C|} \cdot \frac{|CE|}{|EA|} = 1 = \frac{|AF|}{|FB|} \cdot \frac{|BD|}{|DC|} \cdot \frac{|CE|}{|EA|},$$

implying that $\frac{|BD_1|}{|D_1C|} = \frac{|BD|}{|DC|}$. Because both D and D_1 lie on segment BC, we conclude that $D = D_1$, establishing part (1).

Using Ceva's theorem, we can see that the medians, altitudes, and angle bisectors of a triangle are concurrent. The names of these concurrent points are the centroid (G), orthocenter (H), and incenter (I), respectively (Figure 1.30). If the incircle of triangle ABC touches sides AB, BC, and CA at F, D, and E, then by equal tangents, we have $|AE| = |AF|$, $|BD| = |BF|$, and $|CD| = |CE|$. By Ceva's theorem, it follows that lines AD, BE, and CF are concurrent, and the point of concurrency is called the **Gergonne point** (Ge) of the triangle. All these four points are shown in Figure 1.30. Given an angle, it is not difficult to see that the points lying on the bisector of the angle are equidistant from the rays forming the angle. Thus, the intersection of the three angle bisectors is equidistant from the three sides. Hence, this intersection point is the center of the unique circle that is inscribed in the triangle. That is why this point is the incenter of the triangle.

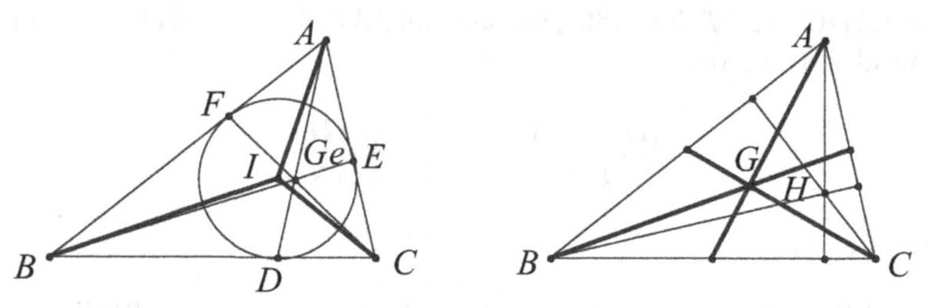

Figure 1.30.

Note that Ceva's theorem can be generalized in a such a way that the point of concurrency does not necessarily have to be inside the triangle; that is, the cevian can be considered as a segment joining a vertex and a point lying on the line of the opposite side. The reader might want to establish the theorem for the configuration shown in Figure 1.31.

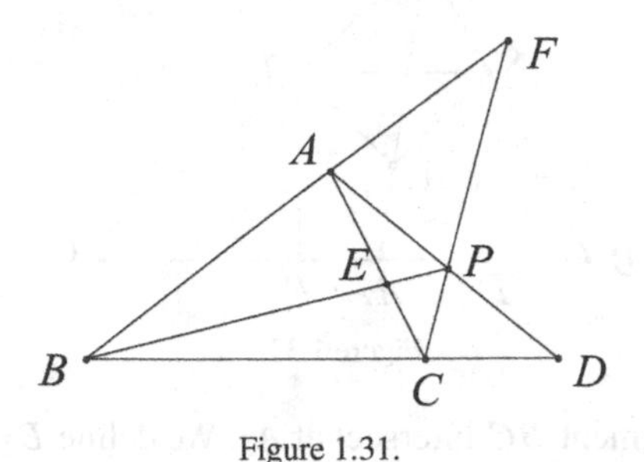

Figure 1.31.

With this general form in mind, it is straightforward to see that in a triangle, the two exterior angle bisectors at two of its vertices and the interior angle bisector at the third vertex are concurrent, and the point of concurrency is the **excenter** of the triangle opposite the third vertex. Figure 1.32 shows the excenter I_A of triangle ABC opposite A. Following the reasoning of the definition of the incenter, it is not difficult to see that I_A is the center of the unique circle outside of triangle ABC that is tangent to rays AB and AC and side BC.

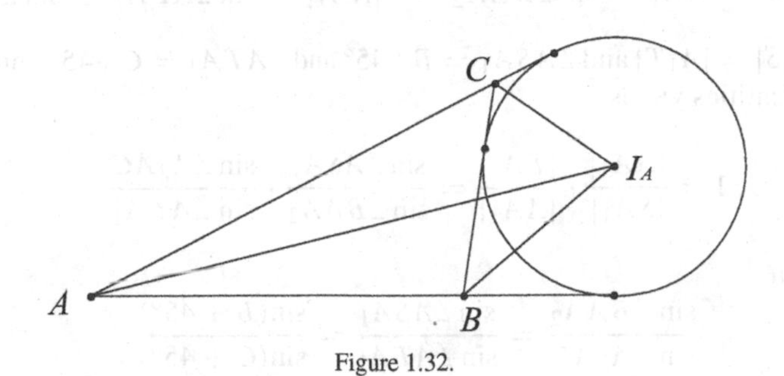

Figure 1.32.

The following example is another good application of Ceva's theorem.

Example 1.12. [IMO 2001 Short List] Let A_1 be the center of the square inscribed in acute triangle ABC with two vertices of the square on side BC (Figure 1.33). Thus one of the two remaining vertices of the square lies on side AB and the other on segment AC. Points B_1 and C_1 are defined in a similar way for inscribed squares with two vertices on sides AC and AB, respectively. Prove that lines AA_1, BB_1, CC_1 are concurrent.

Solution:

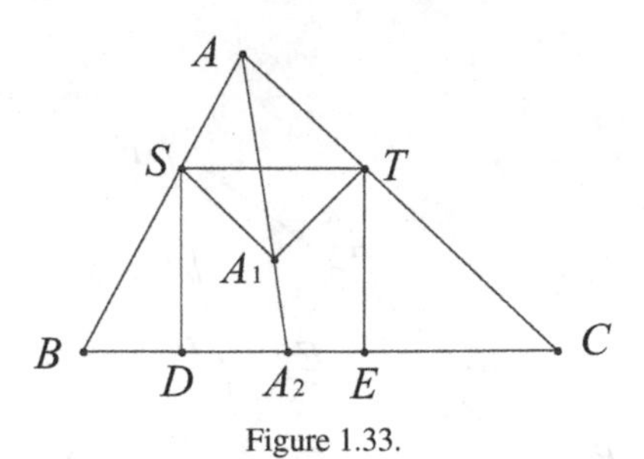

Figure 1.33.

Let line AA_1 and segment BC intersect at A_2. We define B_2 and C_2 analogously. By Ceva's theorem, it suffices to show that

$$\frac{\sin \angle BAA_2}{\sin \angle A_2AC} \cdot \frac{\sin \angle CBB_2}{\sin \angle B_2BA} \cdot \frac{\sin \angle ACC_2}{\sin \angle C_2CB} = 1.$$

Let the vertices of the square be $DETS$, labled as shown in Figure 1.33. Applying the **law of sines** to triangles ASA_1 and ATA_1 gives

$$\frac{|AA_1|}{|SA_1|} = \frac{\sin \angle ASA_1}{\sin \angle SAA_1} = \frac{\sin \angle ASA_1}{\sin \angle BAA_2} \text{ and } \frac{|TA_1|}{|AA_1|} = \frac{\sin \angle A_1AT}{\sin \angle ATA_1} = \frac{\sin \angle A_2AC}{\sin \angle ATA_1}.$$

Because $|A_1S| = |A_1T|$ and $\angle ASA_1 = B+45°$ and $\angle ATA_1 = C+45°$, multiplying the above identities yields

$$1 = \frac{|AA_1|}{|SA_1|} \cdot \frac{|TA_1|}{|AA_1|} = \frac{\sin \angle ASA_1}{\sin \angle BAA_2} \cdot \frac{\sin \angle A_2AC}{\sin \angle ATA_1},$$

implying that

$$\frac{\sin \angle BAA_2}{\sin \angle A_2AC} = \frac{\sin \angle ASA_1}{\sin \angle ATA_1} = \frac{\sin(B+45°)}{\sin(C+45°)}.$$

In exactly the same way, we can show that

$$\frac{\sin \angle CBB_2}{\sin \angle B_2BA} = \frac{\sin(C+45°)}{\sin(A+45°)} \text{ and } \frac{\sin \angle ACC_2}{\sin \angle C_2CB} = \frac{\sin(A+45°)}{\sin(B+45°)}.$$

Multiplying the last three identities establishes the desired result. ■

Naturally, we can ask the following question: Given a triangle ABC, how does one construct, using only a compass and straightedge, a square DEE_1D_1 inscribed in triangle ABC with D and E on side BC?

Think Outside the Box

A **homothety** (or **central similarity**, or **dilation**) is a transformation that fixes one point O (called its **center**) and maps each point P to a point P' for which O, P, and P' are collinear and the ratio $OP : OP' = k$ is constant (k can be either positive or negative). The constant k is called the **magnitude** of the homothety. The point P' is called the **image** of P, and P the **preimage** of P'.

We can now answer our previous question. As shown in Figure 1.34, we first construct a square BCE_2D_2 outside of triangle ABC. (With compass and straightedge, it is possible to construct a line perpendicular to a given line. How?) Let lines AD_2 and AE_2 meet segment BC at D and E, respectively. Then we claim that D and E are two of the vertices of the square that we are looking for. Why? If line D_2E_2 intersects lines AB and AC at B_2 and C_2, then triangles ABC and AB_2C_2 are homothetic (with center A); that is, there is a dilation centered at A that takes triangle ABC, point by point, to triangle AB_2C_2. It is not difficult to see that the magnitude of the homothety is $\frac{|AB_2|}{|AB|} = \frac{|AC_2|}{|AC|} = \frac{|B_2C_2|}{|BC|}$. Note that square BCE_2D_2 is inscribed in triangle AB_2C_2. Hence D and E, the preimages of D_2 and E_2, are the two desired vertices of the inscribed square of triangle ABC.

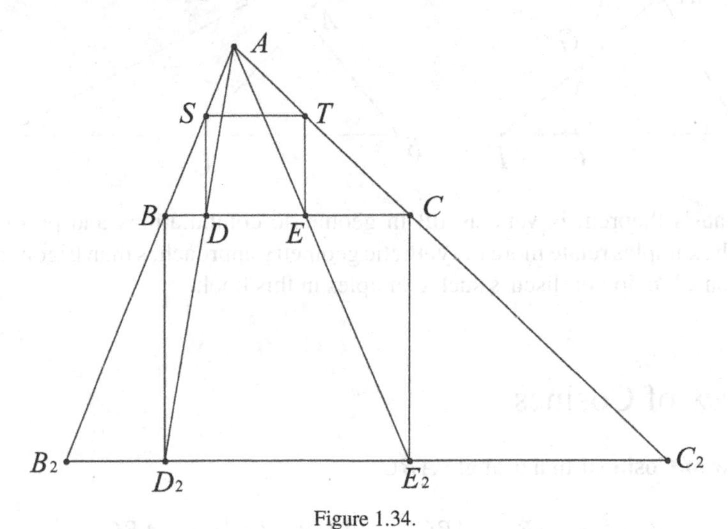

Figure 1.34.

Menelaus's Theorem

While **Ceva's theorem** concerns the concurrency of lines, **Menelaus's theorem** is about the collinearity of points.

[**Menelaus's Theorem**] Given a triangle ABC, let F, G, H be points on lines BC, CA, AB, respectively (Figure 1.35). Then F, G, H are collinear if and only if

$$\frac{|AH|}{|HB|} \cdot \frac{|BF|}{|FC|} \cdot \frac{|CG|}{|GA|} = 1.$$

This is yet another application of the **law of sines**. Applying the law of sines to triangles AGH, BFH, and CFG yields

$$\frac{|AH|}{|GA|} = \frac{\sin \angle AGH}{\sin \angle GHA}, \quad \frac{|BF|}{|HB|} = \frac{\sin \angle BHF}{\sin \angle HFB}, \quad \frac{|CG|}{|FC|} = \frac{\sin \angle GFC}{\sin \angle CGF}.$$

Multiplying the last three identities gives the desired result. (Note that $\sin \angle AGH = \sin \angle CGF$, $\sin \angle BHF = \sin \angle GHA$, and $\sin \angle GFC = \sin \angle HFB$.)

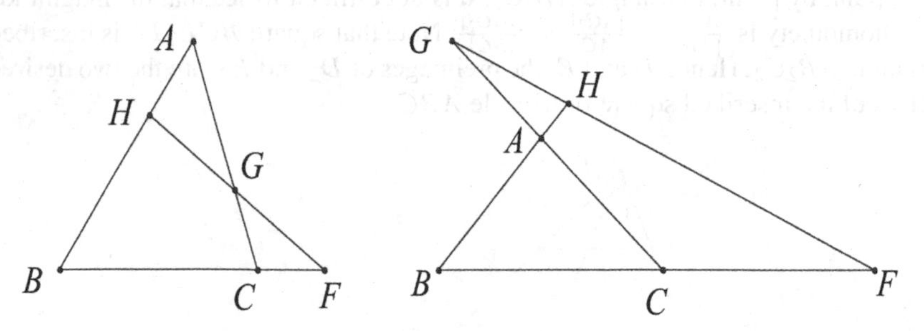

Menelaus's theorem is very useful in geometric computations and proofs. But most such examples relate more to synthetic geometry approaches than trigonometric calculations. We do not discuss such examples in this book.

The Law of Cosines

[**The Law of Cosines**] In a triangle ABC,

$$|CA|^2 = |AB|^2 + |BC|^2 - 2|AB| \cdot |BC| \cos \angle ABC,$$

or, using standard notation,

$$b^2 = c^2 + a^2 - 2ca \cos B,$$

and analogous equations hold for $|AB|^2$ and $|BC|^2$.

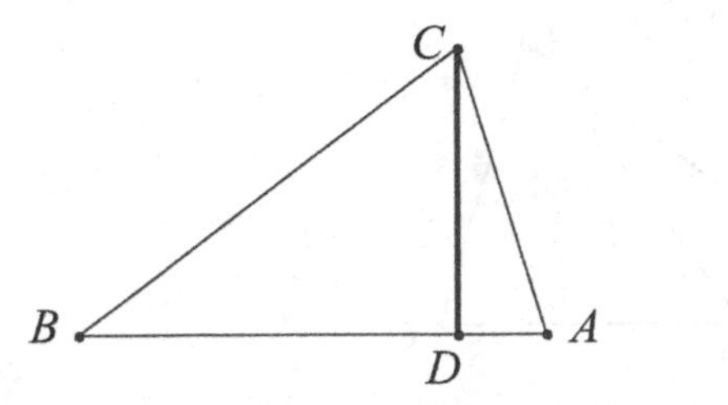

Figure 1.36.

Indeed, let D be the foot of the perpendicular line segment from C to the line AB (Figure 1.36). Then in the right triangle BCD, $|BD| = a\cos B$ and $|CD| = a\sin B$. Hence, $|DA| = |c - a\cos B|$; here we consider the two configurations $0° < A \le 90°$ and $90° < A < 180°$. Then in the right triangle ACD, we have

$$\begin{aligned} b^2 = |CA|^2 = |CD|^2 + |AD|^2 &= a^2\sin^2 B + (c - a\cos B)^2 \\ &= a^2\sin^2 B + c^2 + a^2\cos^2 B - 2ac\cos B \\ &= c^2 + a^2 - 2ca\cos B, \end{aligned}$$

by noting that $\sin^2 B + \cos^2 B = 1$.

From the length of side AB, (including) angle $\angle ABC$, and the length of side BC, by the law of cosines, we can compute the length of the third side BC. This is called the SAS (side–angle–side) form of the law of cosines. On the other hand, solving for $\cos\angle ABC$ gives

$$\cos\angle ABC = \frac{|AB|^2 + |BC|^2 - |CA|^2}{2|AB|\cdot|BC|},$$

or

$$\cos B = \frac{c^2 + a^2 - b^2}{2ca},$$

and analogous formulas hold for $\cos C$ and $\cos A$. This is the SSS (side–side–side) form of the law of cosines.

The Law of Cosines in Action, Take I: Stewart's Theorem

[**Stewart's Theorem**] Let ABC be a triangle, and let D be a point on $\overline{BC}$ (Figure 1.37). Then

$$|BC|\left(|AD|^2 + |BD|\cdot|CD|\right) = |AB|^2\cdot|CD| + |AC|^2\cdot|BD|.$$

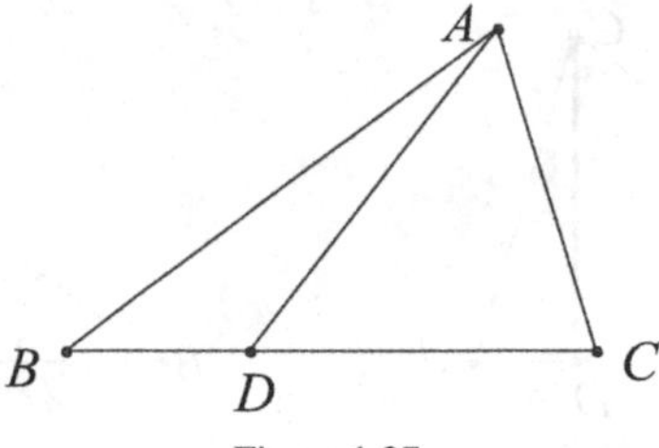

Figure 1.37.

We apply the **law of cosines** to triangles ABD and ACD to obtain

$$\cos\angle ADB = \frac{|AD|^2 + |BD|^2 - |AB|^2}{2|AD|\cdot|BD|}$$

and

$$\cos\angle ADC = \frac{|AD|^2 + |CD|^2 - |AC|^2}{2|AD|\cdot|CD|}.$$

Because $\angle ADB + \angle ADC = 180^\circ$, $\cos\angle ADB + \cos\angle ADC = 0$; that is,

$$\frac{|AD|^2 + |BD|^2 - |AB|^2}{2|AD|\cdot|BD|} + \frac{|AD|^2 + |CD|^2 - |AC|^2}{2|AD|\cdot|CD|} = 0.$$

Multiplying $2AD\cdot BD\cdot CD$ on both sides of the last equation gives

$$|CD|\left(|AD|^2 + |BD|^2 - |AB|^2\right) + |BD|\left(|AD|^2 + |CD|^2 - |AC|^2\right) = 0,$$

or

$$\begin{aligned}
&|AB|^2\cdot|CD| + |AC|^2\cdot|BD| \\
&= |CD|\left(|AD|^2 + |BD|^2\right) + |BD|\left(|AD|^2 + |CD|^2\right) \\
&= (|CD| + |BD|)|AD|^2 + |BD|\cdot|CD|(|BD| + |CD|) \\
&= |BC|(|AD|^2 + |BD|\cdot|CD|).
\end{aligned}$$

Setting $D = M$, the midpoint of segment BC, we can compute the length of the median AM as a special case of Stewart's theorem. We have

$$|AB|^2\cdot|CM| + |AC|^2\cdot|BM| = |BM|(|AM|^2 + |BM|\cdot|CM|),$$

or

$$c^2\cdot\frac{a}{2} + b^2\cdot\frac{a}{2} = a\left(|AM|^2 + \frac{a}{2}\cdot\frac{a}{2}\right).$$

It follows that $2c^2 + 2b^2 = 4|AM|^2 + a^2$, or

$$|AM|^2 = \frac{2b^2 + 2c^2 - a^2}{4},$$

which is the **median formula**.

The Law of Cosines in Action, Take II: Heron's Formula and Brahmagupta's Formula

[**Brahmagupta's Formula**] Let $ABCD$ be a convex cyclic quadrilateral (Figure 1.38). Let $|AB| = a$, $|BC| = b$, $|CD| = c$, $|DA| = d$, and $s = (a+b+c+d)/2$. Then

$$[ABCD] = \sqrt{(s-a)(s-b)(s-c)(s-d)}.$$

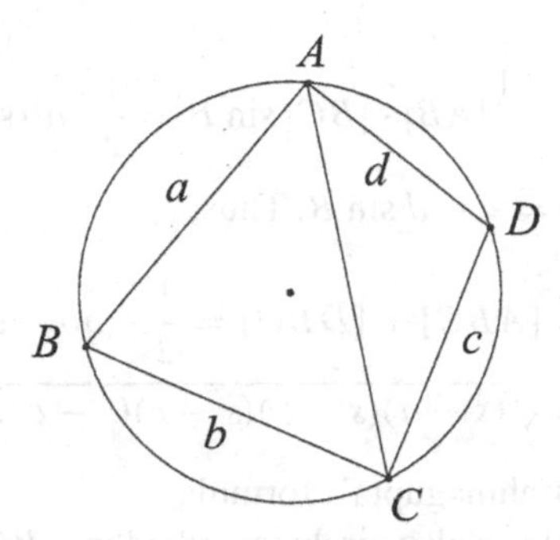

Figure 1.38.

Let $B = \angle ABC$ and $D = \angle ADC$. Applying the **law of cosines** to triangles ABC and DBC yields

$$a^2 + b^2 - 2ab\cos B = AC^2 = c^2 + d^2 - 2cd\cos D.$$

Because $ABCD$ is cyclic, $B + D = 180^\circ$, and so $\cos B = -\cos D$. Hence

$$\cos B = \frac{a^2+b^2-c^2-d^2}{2(ab+cd)}.$$

It follows that

$$\begin{aligned}\sin^2 B &= 1-\cos^2 B = (1+\cos B)(1-\cos B)\\ &= \left(1+\frac{a^2+b^2-c^2-d^2}{2(ab+cd)}\right)\left(1-\frac{a^2+b^2-c^2-d^2}{2(ab+cd)}\right)\\ &= \frac{a^2+b^2+2ab-(c^2+d^2-2cd)}{2(ab+cd)}\cdot\frac{c^2+d^2+2cd-(a^2+b^2-2ab)}{2(ab+cd)}\\ &= \frac{[(a+b)^2-(c-d)^2][(c+d)^2-(a-b)^2]}{4(ab+cd)^2}.\end{aligned}$$

Note that

$$(a+b)^2-(c-d)^2 = (a+b+c-d)(a+b+d-c) = 4(s-d)(s-c).$$

Likewise, $(c+d)^2 - (a-b)^2 = 4(s-a)(s-b)$. Therefore, by observing that $B + D = 180°$ and $0° < B, D < 180°$, we obtain

$$\sin B = \sin D = \frac{2\sqrt{(s-a)(s-b)(s-c)(s-d)}}{ab+cd},$$

or

$$\frac{1}{2} \cdot (ab+cd) \sin B = \sqrt{(s-a)(s-b)(s-c)(s-d)}.$$

Now,

$$[ABC] = \frac{1}{2}|AB| \cdot |BC| \sin B = \frac{1}{2} \cdot ab \sin B.$$

Likewise, we have $[DBC] = \frac{1}{2} \cdot cd \sin B$. Thus,

$$[ABCD] = [ABC] + [DBC] = \frac{1}{2} \cdot (ab+cd) \sin B$$
$$= \sqrt{(s-a)(s-b)(s-c)(s-d)}.$$

This completes the proof Brahmagupta's formula.

Further assume that there is a also circle inscribed in $ABCD$ (Figure 1.39). Then by equal tangents, we have $a + c = b + d = s$, and so $[ABCD] = \sqrt{abcd}$.

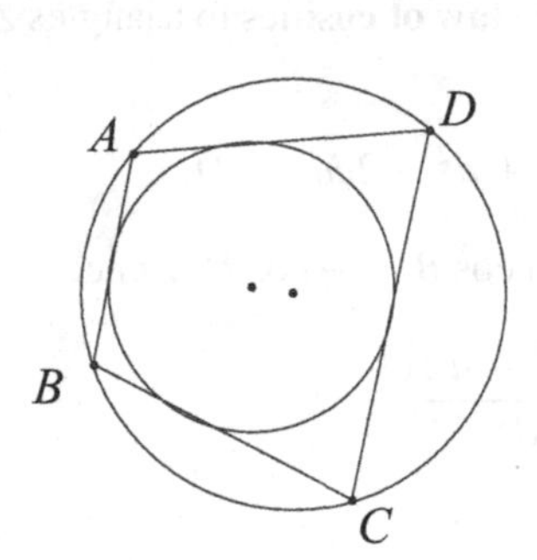

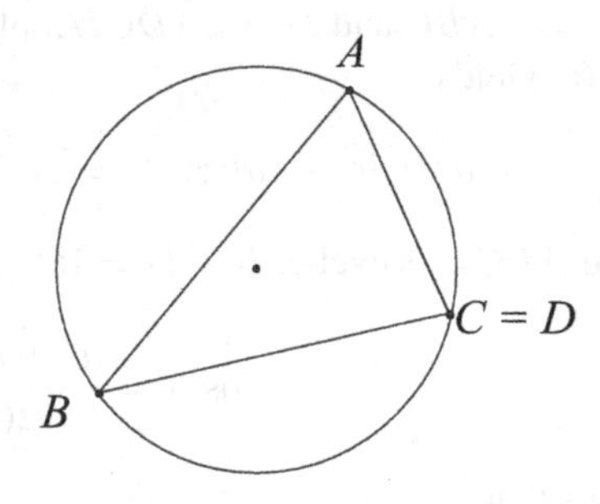

Figure 1.39.

[**Heron's Formula**] The area of a triangle ABC with sides a, b, c is equal to

$$[ABC] = \sqrt{s(s-a)(s-b)(s-c)},$$

where $s = (a+b+c)/2$ is the semiperimeter of the triangle.

Heron's formula can be viewed as a degenerate version of Brahmagupta's formula. Because a triangle is always cyclic, we can view triangle ABC as a cyclic quadrilateral $ABCD$ with $C = D$ (Figure 1.39, right); that is, $CD = 0$. In this way, Brahmagupta's formula becomes Heron's formula. For the interested reader, it is a good exercise to prove Heron's formula independently, following the proof of Brahmagupta's formula.

The Law of Cosines in Action, Take III: Brocard Points

We will show that inside any triangle ABC, there exists a unique point P (Figure 1.40) such that

$$\angle PAB = \angle PBC = \angle PCA.$$

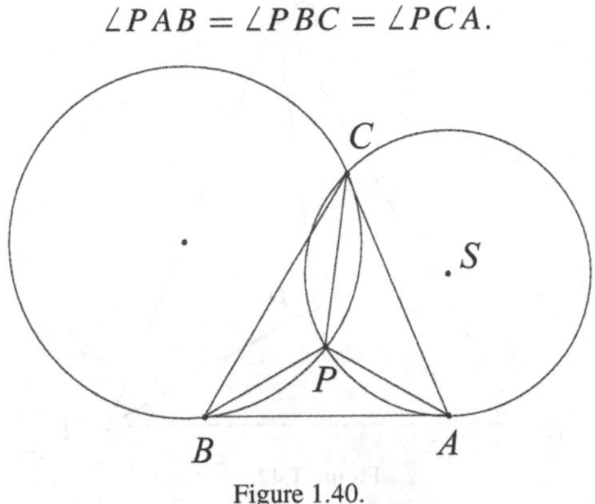

Figure 1.40.

This point is called one of the two **Brocard points** of triangle ABC; the other satisfies similar relations with the vertices in reverse order. Indeed, if $\angle PAB = \angle PCA$, then the circumcircle of triangle ACP is tangent to the line AB at A. If S is the center of this circle, then S lies on the perpendicular bisector of segments AC, and the line SA is perpendicular to the line AB. Hence, this center can be constructed easily. Therefore, point P lies on the circle centered at S with radius $|SA|$ (note that this circle is not tangent to line BC unless $|BA| = |BC|$). We can use the equation $\angle PBC = \angle PCA$ to construct the circle passing through B and tangent to line AC at C. The Brocard point P must lie on both circles and be different from C. Such a point is unique. The third equation $\angle PAB = \angle PBC$ clearly holds.

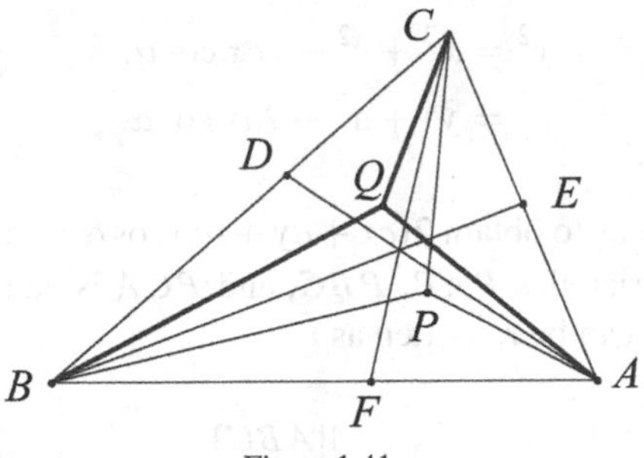

Figure 1.41.

We can construct the other Brocard point in a similar fashion, but in reverse order. We can also reflect lines AP, BP, and CP across the angle bisectors of $\angle CAB$, $\angle ABC$, and $\angle BCA$, respectively (Figure 1.41). Then by **Ceva's theorem**, these three new lines are also concurrent, and the point of concurrency is the second Brocard point. This is the reason we say that the two Brocard points are **isogonal conjugates** of each other.

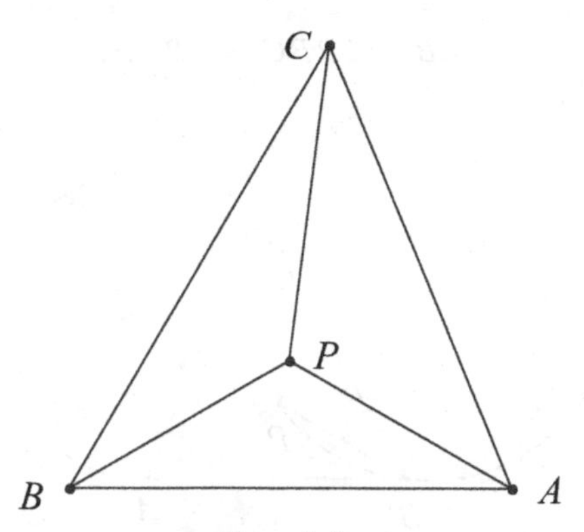

Figure 1.42.

Example 1.13. [AIME 1999] Point P is located inside triangle ABC (Figure 1.42) so that angles PAB, PBC, and PCA are all congruent. The sides of the triangle have lengths $|AB| = 13$, $|BC| = 14$, and $|CA| = 15$, and the tangent of angle PAB is m/n, where m and n are relatively prime positive integers. Find $m + n$.

Solution: Let $\alpha = \angle PAB = \angle PBC = \angle PCA$ and let x, y, and z denote $|PA|$, $|PB|$, and $|PC|$. Apply the **law of cosines** to triangles PCA, PAB, and PBC to obtain

$$x^2 = z^2 + b^2 - 2bz\cos\alpha,$$
$$y^2 = x^2 + c^2 - 2cx\cos\alpha,$$
$$z^2 = y^2 + a^2 - 2ay\cos\alpha.$$

Sum these three equations to obtain $2(cx + ay + bz)\cos\alpha = a^2 + b^2 + c^2$. Because the combined area of triangles PAB, PBC, and PCA is equal to $\frac{(cx+ay+bz)\sin\alpha}{2}$, the preceding equation can be rewritten as

$$\tan\alpha = \frac{4[ABC]}{a^2 + b^2 + c^2}.$$

With $a = 14$, $b = 15$, and $c = 13$, use **Heron's formula** to find that $[ABC] = 84$. It follows that $\tan\alpha = \frac{168}{295}$, so $m + n = 463$. ∎

In general, because $4[ABC] = 2ab \sin C = 2bc \sin A = 2ca \sin B$, we have

$$\begin{aligned}\cot\alpha &= \frac{1}{\tan\alpha} = \frac{a^2+b^2+c^2}{4[ABC]} = \frac{a^2}{2bc\sin A} + \frac{b^2}{2ca\sin B} + \frac{c^2}{2ab\sin C}\\ &= \frac{\sin^2 A}{2\sin B\sin C\sin A} + \frac{\sin^2 B}{2\sin C\sin A\sin B} + \frac{\sin^2 C}{2\sin A\sin B\sin C}\\ &= \frac{\sin^2 A + \sin^2 B + \sin^2 C}{2\sin A\sin B\sin C},\end{aligned}$$

by the **law of sines**.

There is another symmetric identity:

$$\csc^2\alpha = \csc^2 A + \csc^2 B + \csc^2 C.$$

Because $\angle PCA + \angle PAC = \angle PAB + \angle PAC = \angle CAB$, it follows that $\angle CPA = 180^\circ - \angle CAB$, and so $\sin \angle CPA = \sin A$. Applying the law of sines to triangle CAP gives

$$\frac{x}{\sin\alpha} = \frac{b}{\sin\angle CPA}, \quad \text{or} \quad x = \frac{b\sin\alpha}{\sin A}.$$

Similarly, by working with triangles ABP and BCP, we obtain $y = \frac{c\sin\alpha}{\sin B}$ and $z = \frac{a\sin\alpha}{\sin C}$. Consequently,

$$\begin{aligned}[CAP] &= \frac{1}{2}zx\sin\angle CPA = \frac{1}{2}\cdot\frac{a\sin\alpha}{\sin C}\cdot\frac{b\sin\alpha}{\sin A}\cdot\sin A\\ &= \frac{ab\sin C}{2}\cdot\frac{\sin^2\alpha}{\sin^2 C} = [ABC]\cdot\frac{\sin^2\alpha}{\sin^2 C}.\end{aligned}$$

Likewise, we have $[ABP] = [ABC]\cdot\frac{\sin^2\alpha}{\sin^2 A}$ and $[BCP] = [ABC]\cdot\frac{\sin^2\alpha}{\sin^2 B}$. Adding the last three equations gives

$$[ABC] = [ABC]\left(\frac{\sin^2\alpha}{\sin^2 C} + \frac{\sin^2\alpha}{\sin^2 A} + \frac{\sin^2\alpha}{\sin^2 B}\right),$$

implying that $\csc^2\alpha = \csc^2 A + \csc^2 B + \csc^2 C$.

Vectors

In the coordinate plane, let $A = (x_1, y_1)$ and $B = (x_2, y_2)$. We define the **vector** $\overrightarrow{AB} = [x_2 - x_1, y_2 - y_1]$, the displacement from A to B. We use a directed segment to denote a vector. We call the starting (or the first) point (in this case, point A) the

tail of the vector, and the ending (or the second) point (B) the **head**. It makes sense to write the vector $\overrightarrow{AC}$ as the **sum** of vectors $\overrightarrow{AB}$ and $\overrightarrow{BC}$, because the composite displacements from A to B and B to C add up to the displacement from A to C. For example, as shown in Figure 1.43, left, with $A = (10, 45)$, $B = (30, 5)$, and $C = (35, 20)$, then $\overrightarrow{AB} = [20, -40]$, $\overrightarrow{BC} = [5, 15]$, and $\overrightarrow{AC} = [25, -25]$.

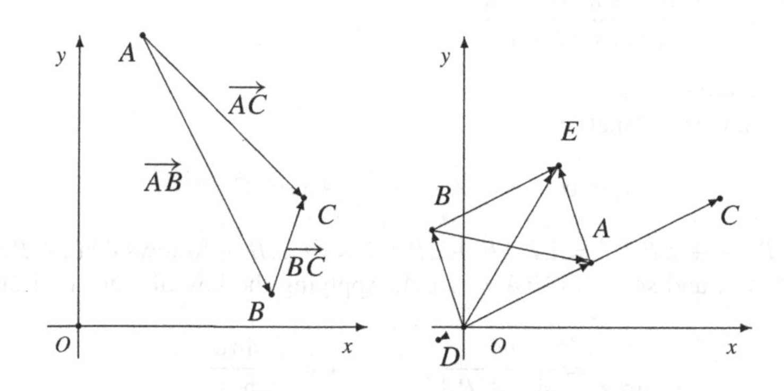

Figure 1.43.

In general (Figure 1.43, right), if $\mathbf{u} = [a, b]$ and $\mathbf{v} = [m, n]$, then $\mathbf{u} + \mathbf{v} = [a+m, b+n]$. If we put the tail of $\mathbf{u}$ at the origin, then its head is at point $A = (a, b)$. If we also put the tail of $\mathbf{v}$ at the origin, then its head is at point $B = (m, n)$. Then $\mathbf{u} + \mathbf{v} = \overrightarrow{OA} + \overrightarrow{OB} = \overrightarrow{OE}$, and $OAEB$ is a parallelogram. We say that vector $\overrightarrow{OA}$ is a **scalar multiple** of $\overrightarrow{OC}$ if there is a constant c such that $\overrightarrow{OC} = [ca, cb]$, and c is called the **scaling factor**. For abbreviation, we also write $[ca, cb]$ as $c[a, b]$. If the vector $\overrightarrow{OA}$ is a scalar multiple of $\overrightarrow{OC}$, it is not difficult to see that O, A, and C are collinear. If c is positive, then we say that the two vectors point in the same direction; if c is negative, we say that the vectors point in opposite directions.

We call $\sqrt{a^2 + b^2}$ the **length** or **magnitude** of vector $\mathbf{u}$, and we denote it by $|\mathbf{u}|$. If $a \neq 0$, we also call $\frac{b}{a}$ the slope of the vector; if $a = 0$, we say that $\mathbf{u}$ is vertical. (These terms are naturally adapted from analytic geometry.) We say that vectors are perpendicular to each other if they form a $90°$ angle when placed tail to tail. By properties of slopes, we can derive that $\mathbf{u}$ and $\mathbf{v}$ are perpendicular if and only if $am + bn = 0$. We can also see this fact by checking that $|OA|^2 + |OB|^2 = |AB|^2$; that is, $|\mathbf{u}|^2 + |\mathbf{v}|^2 = |\mathbf{u} - \mathbf{v}|^2$. It follows that $\mathbf{u}$ and $\mathbf{v}$ are perpendicular if and only if $(a^2 + b^2) + (m^2 + n^2) = (a - m)^2 + (b - n)^2$, or $am + bn = 0$ (Figure 1.44, left).

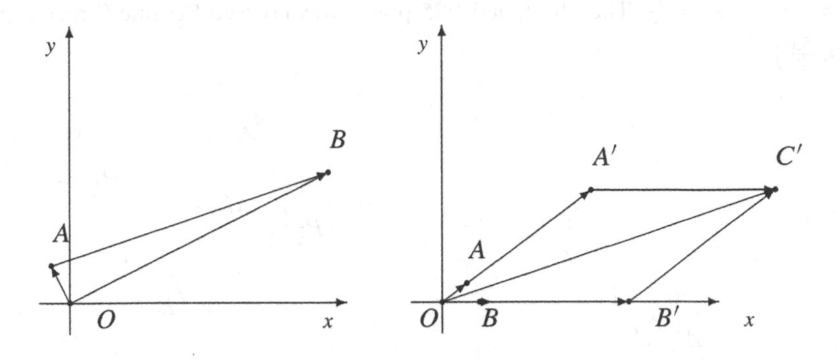

Figure 1.44.

Note that the diagonals of a rhombus bisect the interior angles (Figure 1.44, right). Because vectors $\overrightarrow{OA'} = |\mathbf{v}|\mathbf{u}$ and $\overrightarrow{OB'} = |\mathbf{u}|\mathbf{v}$ have the same length, we note that if vectors $\overrightarrow{OA'}$, $\overrightarrow{OB'}$, and $\overrightarrow{OC'} = \overrightarrow{OA'} + \overrightarrow{OB'} = |\mathbf{v}|\mathbf{u} + |\mathbf{u}|\mathbf{v}$ are placed tail to tail, $\overrightarrow{OC'}$ bisects the angle formed by vectors $\overrightarrow{OA'}$ and $\overrightarrow{OB'}$, which is the same as the angle formed by vectors **u** and **v**.

A vector contains two major pieces of information: its length and its direction (slope). Hence vectors are a very powerful tool for dealing with problems in analytic geometry. Let's see some examples.

Example 1.14. Alex started to wander in Wonderland at 11:00 a.m. At 12:00 p.m., Alex was at $A = (5, 26)$; at 1:00 p.m., Alex was spotted at $B = (-7, 6)$. If Alex moves along a fixed direction at a constant rate, where was Alex at 12:35 p.m.? 11:45 a.m.? 1:30 p.m.? At what time and at what location did Alex cross Sesame Street, the y axis?

Solution: As shown in Figure 1.45, left, let A_3, A_1, A_4 denote Alex's positions at 12:35 p.m., 11:45 p.m., and 1:30 p.m., respectively. It took Alex 60 minutes to move along the vector $\overrightarrow{AB} = [-12, -20]$. Hence he was dislocated by vector $\overrightarrow{AA_3} = \frac{35}{60}\overrightarrow{AB} = \left[-7, -\frac{35}{3}\right]$ at 12:35 p.m. Similarly, $\overrightarrow{AA_1} = -\frac{15}{60}\overrightarrow{AB} = [3, 5]$ and $\overrightarrow{AA_4} = \frac{90}{60}\overrightarrow{AB} = [-18, -30]$. Let $O = (0, 0)$ be the origin. Then we find that $\overrightarrow{OA_3} = \overrightarrow{OA} + \overrightarrow{AA_3} = \left[-2, \frac{43}{3}\right]$ and $A_3 = \left(-2, \frac{43}{3}\right)$. Likewise, $A_1 = (8, 31)$ and $A_4 = (-13, -4)$.

Let $A_2 = (0, b)$ denote the point at which Alex crosses Sesame Street. Assume that it took t minutes after 12:00 p.m. for Alex to cross Sesame Street. Then $\overrightarrow{AA_2} = \frac{t}{60}\overrightarrow{OA_1}$ and $\overrightarrow{OA_2} = \overrightarrow{OA} + \overrightarrow{AA_2}$; that is, $[0, b] = [5, 26] + \frac{t}{60}[-12, -20]$. It follows that $[0, b] = \left[5 - \frac{t}{5}, 26 - \frac{t}{3}\right]$. Solving $0 = 5 - \frac{t}{5}$ gives $t = 25$, which implies that

$b = 26 - \frac{25}{3} = \frac{53}{3}$. Therefore, at 12:25 p.m., Alex crossed Sesame Street at point $\left(0, \frac{53}{3}\right)$. ■

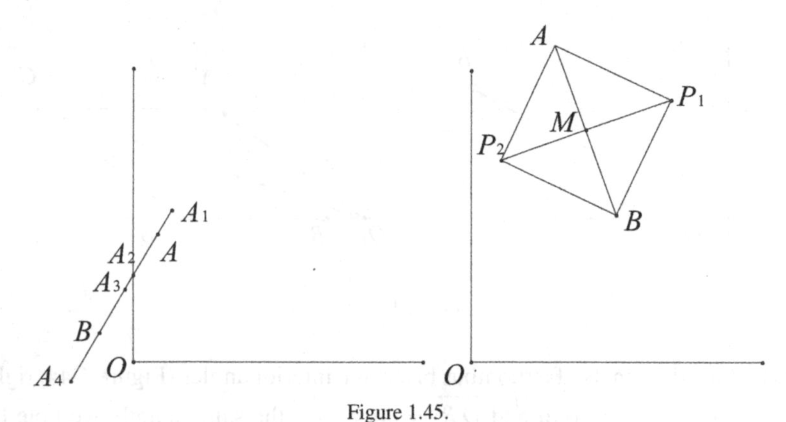

Figure 1.45.

Example 1.15. Given points $A = (7, 26)$ and $B = (12, 12)$, find all points P such that $|AP| = |BP|$ and $\angle APB = 90^\circ$ (Figure 1.45, right).

Solution: Note that triangle ABP is an isosceles right triangle with $|AP| = |BP|$. Let M be the midpoint of segment AB. Then $M = \left(\frac{19}{2}, 19\right)$, $|MA| = |MB| = |MP|$, and $MA \perp MB$. Thus $\overrightarrow{MA} = \left[\frac{5}{2}, -7\right]$, and $\overrightarrow{MP} = \left[7, \frac{5}{2}\right]$ or $\overrightarrow{MP} = -\left[7, \frac{5}{2}\right]$. Let $O = (0, 0)$ be the origin. It follows that $\overrightarrow{OP} = \overrightarrow{OM} + \overrightarrow{MP} = \left[\frac{19}{2}, 19\right] \pm \left[7, \frac{5}{2}\right]$. Consequently, $P = \left(\frac{33}{2}, \frac{43}{2}\right)$ or $P = \left(\frac{5}{2}, \frac{33}{2}\right)$. ■

For each of the next two examples, we present two solutions. The first solution applies vector operations. The second solution applies trigonometric computations.

Example 1.16. [ARML 2002] Starting at the origin, a beam of light hits a mirror (in the form of a line) at point $A = (4, 8)$ and is reflected to point $B = (8, 12)$. Compute the exact slope of the mirror.

Note: The key fact in this problem is the fact that the angle of incidence is equal to the angle of reflection; that is, if the mirror lies on line PQ, as shown in Figure 1.46, left, then $\angle OAQ = \angle PAB$.

First Solution: Construct line ℓ such that $\ell \perp PQ$. Then line ℓ bisects angle $\angle OAB$. Note that $|\overrightarrow{AB}| = \sqrt{(8-4)^2 + (12-8)^2} = 4\sqrt{2}$ and $|\overrightarrow{AO}| = \sqrt{4^2 + 8^2} = 4\sqrt{5}$. Thus, vector $\sqrt{5} \cdot \overrightarrow{AB} + \sqrt{2} \cdot \overrightarrow{AO}$ bisects the angle formed by $\overrightarrow{AO}$ and $\overrightarrow{AB}$; that is, this vector and line ℓ have the same slope. Because $\sqrt{5} \cdot \overrightarrow{AB} + \sqrt{2} \cdot \overrightarrow{AO} =$

$\left[4\sqrt{5}-4\sqrt{2}, 4\sqrt{5}-8\sqrt{2}\right]$, the slope of line ℓ is $\frac{4\sqrt{5}-8\sqrt{2}}{4\sqrt{5}-4\sqrt{2}} = \frac{\sqrt{5}-2\sqrt{2}}{\sqrt{5}-\sqrt{2}}$, and so the slope of line PQ is

$$\frac{\sqrt{5}-\sqrt{2}}{2\sqrt{2}-\sqrt{5}} = \frac{(\sqrt{5}-\sqrt{2})(2\sqrt{2}+\sqrt{5})}{(2\sqrt{2}-\sqrt{5})(2\sqrt{2}+\sqrt{5})} = \frac{\sqrt{10}+1}{3}.$$

Second Solution: Let ℓ_1 and ℓ_2 denote two lines, and for $i = 1$ and 2, let m_i and θ_i (with $0° \le \theta_i < 180°$) denote the slope and the polar angle of line ℓ_i, respectively. Without loss of generality, we assume that $\theta_1 > \theta_2$. If θ is the polar angle formed by the lines, then $\theta = \theta_1 - \theta_2$ and

$$\tan\theta = \frac{\tan\theta_1 - \tan\theta_2}{1+\tan\theta_1\theta_2} = \frac{m_1 - m_2}{1 - m_1 m_2}$$

by the **addition and subtraction formulas.**

Let m be the slope of line PQ. Because lines OA and AB have slopes 2 and 1, respectively, by our earlier discussion we have

$$\frac{m-1}{1+m} = \tan\angle PAB = \tan\angle QAO = \frac{2-m}{1+2m},$$

or $(m-1)(1+2m) = (2-m)(1+m)$. It follows that $3m^2 - 2m - 1 = 0$, or $m = \frac{1\pm\sqrt{10}}{3}$. It is not difficult to see that $m = \frac{1+\sqrt{10}}{3}$ is the answer to the problem. (The other value is the slope of line ℓ, the interior bisector of $\angle OAB$.) ■

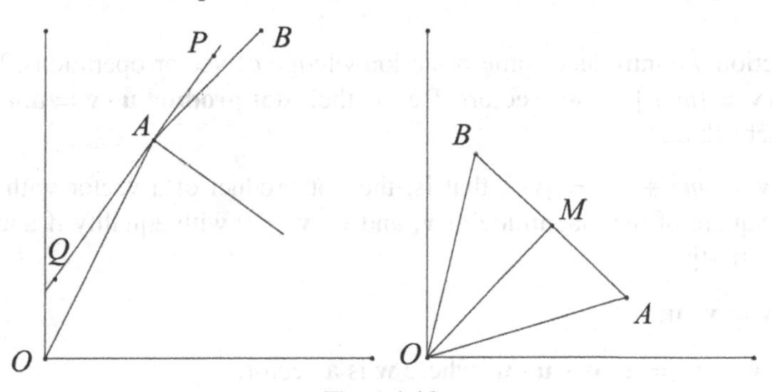

Figure 1.46.

Example 1.17. [AIME 1994] The points $(0, 0)$, $(a, 11)$, and $(b, 37)$ are the vertices of an equilateral triangle (Figure 1.46, right). Find ab.

In both solutions, we set $O = (0, 0)$, $A = (a, 11)$, and $B = (b, 37)$.

First Solution: Let M be the midpoint of segment AB. Then $M = \left(\frac{a+b}{2}, 24\right)$, $OM \perp MA$, and $|OM| = \sqrt{3}|MA|$. Because $\overrightarrow{AM} = \left(\frac{a-b}{2}, -13\right)$, it follows that

$\overrightarrow{OM} = \sqrt{3}\left[13, \frac{a-b}{2}\right]$. Hence, $\left[\frac{a+b}{2}, 24\right] = \sqrt{3}\left[13, \frac{a-b}{2}\right]$; that is,

$$\frac{a+b}{2} = 13\sqrt{3} \quad \text{and} \quad \frac{a-b}{2} = 8\sqrt{3}.$$

Adding the last two equations gives $a = 21\sqrt{3}$, and subtracting the second equation from the first equation gives $b = 5\sqrt{3}$. Consequently, $ab = 315$. ■

Second Solution: Let $\angle\alpha$ denote the angle formed by ray OA and the positive direction of the x axis, and set $x = |OA| = |OB| = |AB|$. Then $\sin\alpha = \frac{11}{x}$ and $\cos\alpha = \frac{a}{x}$. Note that ray OB forms an angle whose measure is $\alpha + 60°$ from the positive x axis. Then by the **addition and subtraction formulas**, we have

$$\frac{37}{x} = \sin(\alpha + 60°) = \sin\alpha\cos 60° + \cos\alpha\sin 60° = \frac{11}{2x} + \frac{a\sqrt{3}}{2x},$$
$$\frac{b}{x} = \cos(\alpha + 60°) = \cos\alpha\cos 60° - \sin\alpha\sin 60° = \frac{a}{2x} - \frac{11\sqrt{3}}{2x}.$$

Solving the first equation for a gives $a = 21\sqrt{3}$. We then solve the second equation for b to obtain $b = 5\sqrt{3}$. Hence $ab = 315$. ■

The Dot Product and the Vector Form of the Law of Cosines

In this section we introduce some basic knowledge of vector operations. Let $\mathbf{u} = [a, b]$ and $\mathbf{v} = [m, n]$ be two vectors. Define their **dot product** $\mathbf{u} \cdot \mathbf{v} = am + bn$. It is easy to check that

(i) $\mathbf{v} \cdot \mathbf{v} = m^2 + n^2 = |\mathbf{v}|^2$; that is, the dot product of a vector with itself is the square of the magnitude of $\mathbf{v}$, and $\mathbf{v} \cdot \mathbf{v} \geq 0$ with equality if and only if $\mathbf{v} = [0, 0]$;

(ii) $\mathbf{u} \cdot \mathbf{v} = \mathbf{v} \cdot \mathbf{u}$;

(iii) $\mathbf{u} \cdot (\mathbf{v} + \mathbf{w}) = \mathbf{u} \cdot \mathbf{v} + \mathbf{u} \cdot \mathbf{w}$, where $\mathbf{w}$ is a vector;

(iv) $(c\mathbf{u}) \cdot \mathbf{v} = c(\mathbf{u} \cdot \mathbf{v})$, where c is a scalar.

If vectors $\mathbf{u}$ and $\mathbf{v}$ are placed tail to tail at the origin O, let A and B be the heads of $\mathbf{u}$ and $\mathbf{v}$, respectively. Then $\overrightarrow{AB} = \mathbf{v} - \mathbf{u}$. Let θ denote the angle formed by lines OA and OB. Then $\angle AOB = \theta$. Applying the **law of cosines** to triangle AOB yields

$$\begin{aligned} |\mathbf{v} - \mathbf{u}|^2 = AB^2 &= OA^2 + OB^2 - 2OA \cdot OB\cos\theta \\ &= |\mathbf{u}|^2 + |\mathbf{v}|^2 - 2|\mathbf{u}||\mathbf{v}|\cos\theta. \end{aligned}$$

It follows that

$$(\mathbf{v}-\mathbf{u})\cdot(\mathbf{v}-\mathbf{u})=\mathbf{u}\cdot\mathbf{u}+\mathbf{v}\cdot\mathbf{v}-2|\mathbf{u}||\mathbf{v}|\cos\theta,$$

or $\mathbf{v}\cdot\mathbf{v}-2\mathbf{u}\cdot\mathbf{v}+\mathbf{u}\cdot\mathbf{u}=\mathbf{v}\cdot\mathbf{v}+\mathbf{u}\cdot\mathbf{u}-2|\mathbf{u}||\mathbf{v}|\cos\theta$. Hence

$$\cos\theta=\frac{\mathbf{u}\cdot\mathbf{v}}{|\mathbf{u}||\mathbf{v}|}.$$

Cauchy–Schwarz Inequality

Let $\mathbf{u}=[a,b]$ and $\mathbf{v}=[m,n]$, and let θ be the angle formed by the two vectors when they are placed tail to tail. Because $|\cos\theta|\le 1$, by the previous discussions, we conclude that $(\mathbf{u}\cdot\mathbf{v})^2\le(|\mathbf{u}||\mathbf{v}|)^2$; that is,

$$(am+bn)^2\le\left(a^2+b^2\right)\left(m^2+n^2\right).$$

Equality holds if and only if $|\cos\theta|=1$, that is, if the two vectors are parallel. In any case, the equality holds if and only if $\mathbf{u}=k\cdot\mathbf{v}$ for some nonzero real constant k; that is, $\frac{a}{m}=\frac{b}{n}=k$.

We can generalize the definitions of vectors for higher dimensions, and define the dot product and the length of the vectors accordingly. This results in **Cauchy–Schwarz inequality**: For any real numbers $a_1, a_2, \ldots, a_n$, and $b_1, b_2, \ldots, b_n$,

$$\left(a_1^2+a_2^2+\cdots+a_n^2\right)\left(b_1^2+b_2^2+\cdots+b_n^2\right)\ge(a_1b_1+a_2b_2+\cdots+a_nb_n)^2.$$

Equality holds if and only if a_i and b_i are proportional, $i=1,2,\ldots,n$.

Now we revisit Example 1.11. Setting $n=3$, $(a_1,a_2,a_3)=\left(\sqrt{x},\sqrt{y},\sqrt{z}\right)$, $(b_1,b_2,b_3)=\left(\frac{1}{\sqrt{x}},\frac{2}{\sqrt{y}},\frac{3}{\sqrt{z}}\right)$ in Cauchy–Schwarz inequality, we have

$$\frac{1}{x}+\frac{4}{y}+\frac{9}{z}=(x+y+z)\left(\frac{1}{x}+\frac{4}{y}+\frac{9}{z}\right)\ge(1+2+3)^2=36.$$

Equality holds if and only if $\frac{x}{1}=\frac{y}{2}=\frac{z}{3}$, or $(x,y,z)=\left(\frac{1}{6},\frac{1}{3},\frac{1}{2}\right)$.

Radians and an Important Limit

When a point moves along the unit circle from $A=(1,0)$ to $B=(0,-1)$, it has traveled a distance of π and through an angle of 180°. We can use the arc length as a way of measuring an angle. We set a conversion between units: $\pi = 180°$; that is, 180° is π **radians.** Hence 1 radian is equal to $\frac{180}{\pi}$ degrees, which is about 57.3°. Therefore, an angle of $\alpha = x°$ has radian measure $x \cdot \frac{\pi}{180}$, and an angle of $\theta = y$ radians has a measure of $y \cdot \frac{180}{\pi}$ degrees. To be a good problem solver, the reader is encouraged to be very familiar with the radian measures of special angles such as 12°, 15°, 30°, 45°, 60°, 120°, 135°, 150°, 210°, and vice versa.

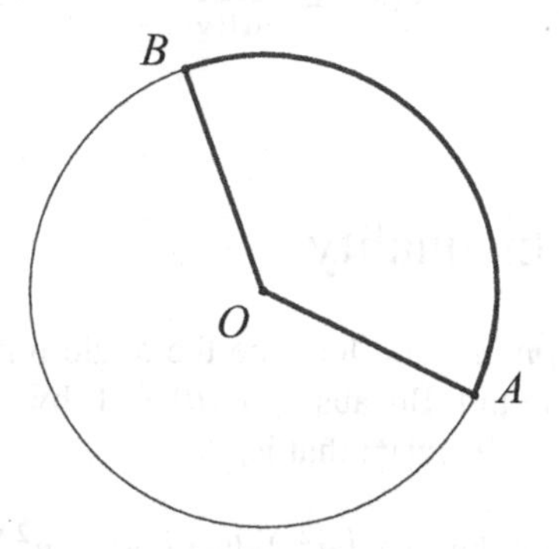

Figure 1.47.

Let ω be a circle, and let O and R denote its center and radius, respectively. Suppose points A and B lie on the circle (Figure 1.47). Assume that $\angle AOB = x°$ and $\angle AOB = y$ (radian measure). Then $\frac{x\pi}{180} = y$. Let $|\widehat{AB}|$ denote the length of arc AB. By the symmetry of the circle, we have

$$\frac{|\widehat{AB}|}{2\pi R} = \frac{x°}{360°}, \quad \text{or} \quad |\widehat{AB}| = \frac{x\pi}{180} \cdot R = yR.$$

Hence, if $\angle AOB$ is given in radian measure, then the length of arc AB is equal to the product of this measure and the radius of the circle. Also, by the symmetry of the circle, the area of this **sector** (the region enclosed by the circle and the two radii OA and OB) is equal to

$$\frac{x°}{360°} \cdot \pi R^2 = \frac{yR^2}{2}.$$

That is, the area of a sector is equal to half of the product of the radian measure of its central angle and the square of the radius of the circle.

Because radian measure reveals an important relation between the size of the central angle of a circle and the arc length that the angle subtends, it is a better unit by which to (algebraically) quantify a geometric object. From now on in this book, for a trigonometric function $f(x)$, we assume that x is in radian measure, unless otherwise specified.

Let θ be a angle with $0 < \theta < \frac{\pi}{2}$. We claim that

$$\sin\theta < \theta < \tan\theta. \qquad (*)$$

We consider the unit circle centered at $O = (0, 0)$, and points $A = (1, 0)$ and $B = (\cos\theta, \sin\theta)$ (Figure 1.48). Then $\angle AOB = \theta$. Let C be the foot of the perpendicular line segment from B to segment OA. Point D lies on ray OB such that $AD \perp AO$. Then $BC = \sin\theta$ and $AD = \tan\theta$. (By our earlier discussion, arc AB has length $1\cdot\theta = \theta$. Hence it is equivalent to show that the lengths of segments BC, arc AB, and segment AD are in increasing order. The reader might also want to find the geometric interpretation of $\cot\theta$, $\sec\theta$, and $\csc\theta$.) It is clear that the areas of triangle OAB, sector OAB, and triangle OAD are in increasing order; that is,

$$\frac{|BC|\cdot|OA|}{2} < \frac{1^2\cdot\theta}{2} < \frac{|AD|\cdot|OA|}{2},$$

from which the desired result follows.

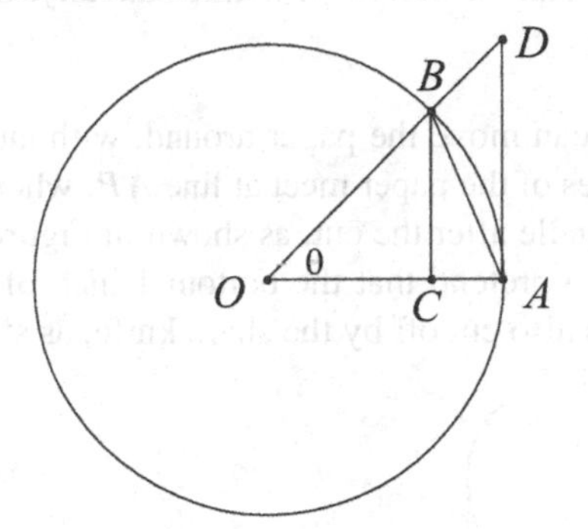

Figure 1.48.

As θ approaches 0, $\sin\theta$ approaches 0. We say that the **limiting value** of $\sin\theta$ is 0 as θ approaches 0, and we denote this fact by $\lim_{\theta\to 0}\sin\theta = 0$. (The reader might want to explain the identity $\lim_{\theta\to 0}\cos\theta = \lim_{\theta\to 0}\sec\theta = 1$.) What about the ratio $\frac{\theta}{\sin\theta}$? Dividing by $\sin\theta$ on all sides of the inequalities in $(*)$ gives

$$1 < \frac{\theta}{\sin\theta} < \frac{1}{\cos\theta} = \sec\theta.$$

Because $\lim\limits_{\theta\to 0}\sec\theta = 1$, it is not difficult to see that the value of $\frac{\theta}{\sin\theta}$, which is sandwiched in between 1 and $\sec\theta$, approaches 1 as well; that is,

$$\lim_{\theta\to 0}\frac{\theta}{\sin\theta} = 1, \quad \text{or} \quad \lim_{\theta\to 0}\frac{\sin\theta}{\theta} = 1.$$

This limit is the foundation of the computations of the **derivatives** of trigonometric functions in calculus.

Note: We were rather vague about the meaning of the term **approaching**. Indeed, when θ approaches 0, it can be either a small positive value or a negative value with small magnitude. These details can be easily dealt with in calculus, which is not the focal point of this book. We introduce this important limit only to illustrate the importance of radian measure.

Constructing Sinusoidal Curves with a Straightedge

Example 1.18. [Phillips Exeter Academy Math Materials] Jackie wraps a sheet of paper tightly around a wax candle whose diameter is two units, then cuts though them both with a sharp knife, making a 45° angle with the candle's axis. After unrolling the paper and laying it flat, Jackie sees the wavy curve formed by the cut edge, and wonders whether it can be described mathematically. Show that this curve is sinusoidal.

Solution: Because we can move the paper around, without loss of generality, we may assume that the sides of the paper meet at line AP, where A is the lowest point on the top face of the candle after the cut, as shown in Figure 1.49. To simplify our work a bit more, we also pretend that the bottom 1 inch of the candle (and paper wrapping around it) was also cut off by the sharp knife, as shown in Figure 1.49.

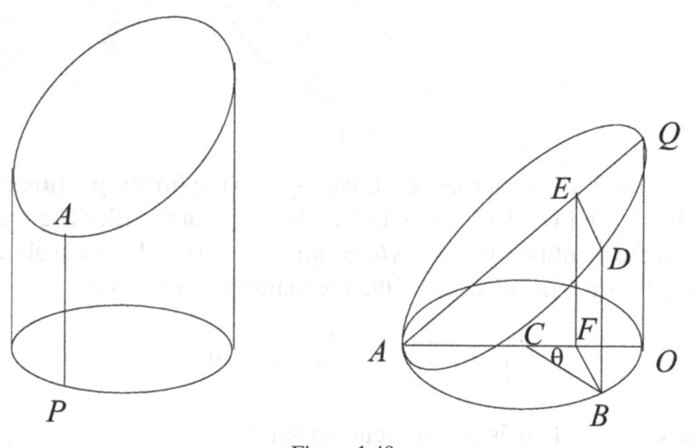

Figure 1.49.

Let $\mathcal{S}$ denote the curve formed by the cut edge before the paper is unrolled, and let D be an arbitrary point on $\mathcal{S}$. For a point X on $\mathcal{S}$, let X_1 denote the point corresponding to F after the paper is unrolled. (When we unroll the paper, point A will correspond to two points A_1 and A_2.) Let O be the point on the base of the candle that is diametrically opposite to A. Let Q be the point on $\mathcal{S}$ such that line QO is parallel to the axis of the candle. Set the coordinate system (on the unrolled paper) in such a way that $O_1 = (0, 0)$, $Q_1 = (0, 2)$, and $A_1 = (-\pi, 0)$. Then by symmetry, $A_2 = (\pi, 0)$. Let C be the midpoint of segment AO, and let ω denote the circle centered at C with radius CA; that is, ω denotes the boundary of the base of the candle.

Let B be the foot of the perpendicular line segment from D to the circle ω, and assume that $\angle OCB = \theta$. Because circle ω has radius 1, $|\widehat{OB}|$, the length of arc OB, is θ. (This is why we use radian measure for θ.) Then $B_1 = (\theta, 0)$ and $D_1 = (\theta, y)$ with $y = BD$. Let F be the foot of the perpendicular line segment from B to segment AC. Then $CF = \cos\theta$, and $AF = 1 + \cos\theta$. Note that A, Q, C, F, and O are coplanar, and $\angle OAQ = 45°$. Point E is on segment AQ such that $EF \perp AO$. Consequently, $\angle AEF = \angle OAQ = 45°$ and $\angle AFE = 90°$, implying that the right triangle AEF is isosceles, with $AF = EF$. It is not difficult to see that $BDEF$ is a rectangle. Hence $BD = EF = AF = 1 + \cos\theta$. We conclude that $D_1 = (\theta, 1 + \cos\theta)$; that is, D_1 lies on the curve $y = 1 + \cos x$.

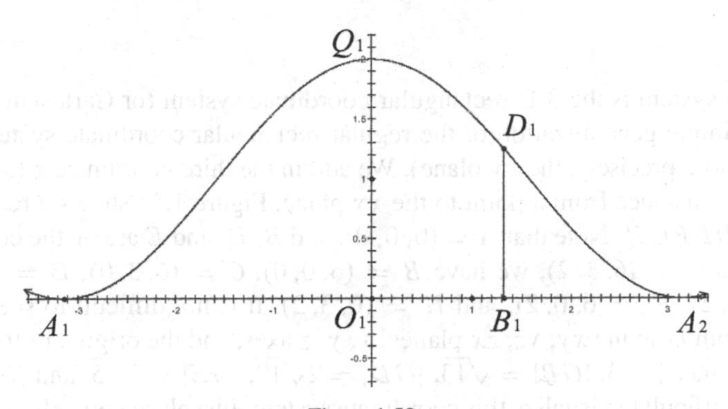

Figure 1.50.

Finally, had we not cut off the bottom of the candle, the equation of the curve would have been $y = 2 + \cos x$. ■

Three Dimensional Coordinate Systems

We view Earth as a sphere, with radius 3960 miles. We will set up two kinds of 3-D coordinate systems to describe the positions of places on Earth.

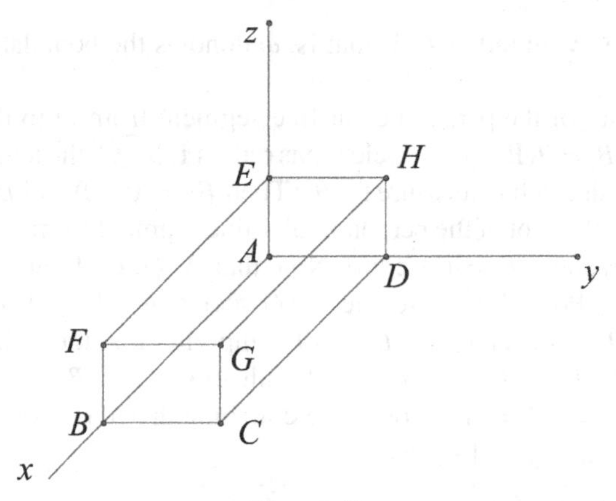

Figure 1.51.

The first system is the 3-D rectangular coordinate system (or Cartesian system). This is a simple generalization of the regular rectangular coordinate system in the plane (or more precisely, the xy plane). We add in the third coordinate z to describe the directed distance from a point to the xy plane. Figure 1.51 shows a rectangular box $ABCDEFGH$. Note that $A = (0, 0, 0)$, and B, D, and E are on the coordinate axes. Given $G = (6, 3, 2)$, we have $B = (6, 0, 0)$, $C = (6, 3, 0)$, $D = (0, 3, 0)$, $E = (0, 0, 2)$, $F = (6, 0, 2)$, and $H = (0, 3, 2)$. It is not difficult to see that the distance from G to the xy, yz, zx planes, x, y, z axes, and the origin are $|GC| = 2$, $|GH| = 6$, $|GF| = 3$, $|GB| = \sqrt{13}$, $|GD| = 2\sqrt{10}$, $|GE| = 3\sqrt{5}$, and $|GA| = 7$.

It is not difficult to visualize this coordinate system. Just place yourself in a regular room, choose a corner on the floor (if you are good at seeing the world upside down, you might want to try a corner on the ceiling) as the origin, and assign the three edges going out of the chosen corner as the three axes. In general, for a point $P = (x, y, z)$, x denotes the directed distance from P to the yz plane, y denotes the directed distance from P to the zx plane, and z denotes the directed distance from P to the xy plane. It is not difficult to see that $\sqrt{x^2 + y^2}$, $\sqrt{y^2 + z^2}$, and $\sqrt{z^2 + x^2}$ are the respective distances from P to the z axis, x axis, and y axis. It is also not difficult to see that the distance between two points $P_1 = (x_1, y_1, z_1)$ and $P_2 = (x_2, y_2, z_2)$ is $\sqrt{(x_1 - x_2)^2 + (y_1 - y_2)^2 + (z_1 - z_2)^2}$. Based on this generalization, we can talk about vectors in 3-D space and their lengths, and the angles formed by them when they are placed tail to tail. (Note that we cannot talk about the standard angle any more.) Hence we can easily generalize the definition of the dot product of three-dimensional vectors $\mathbf{u} = [a, b, c]$ and $\mathbf{v} = [m, n, p]$ as $\mathbf{u} \cdot \mathbf{v} = am + bn + cp$, and it is routine to check that all the properties of the dot product discussed earlier hold.

Let O be the center of Earth. We set the plane containing the equator as the xy plane (or **equatorial plane**), and let the North Pole lie on the positive z axis. However, sometimes it is not convenient to use only the rectangular system to describe positions of places on Earth, simply because Earth is a sphere.

Example 1.19. Describe the points on Earth's surface that can be seen from a space station that is 100 miles above the North Pole.

Solution: Let S denote the position of the space station, and let E be a point on Earth such that line SE is tangent to Earth's surface. There are many such points E, and all these points form a circle $\mathcal{C}$ lying in a plane $\mathcal{P}$ that is parallel to the equatorial plane. The plane $\mathcal{P}$ cuts Earth's surface into two parts. The part containing the North Pole contains exactly those points we are looking for. The best way to describe the points on $\mathcal{C}$ is to use the angle formed by ray OE and the equatorial plane. It is not difficult to see that as E varies along $\mathcal{C}$, the angle does not change.

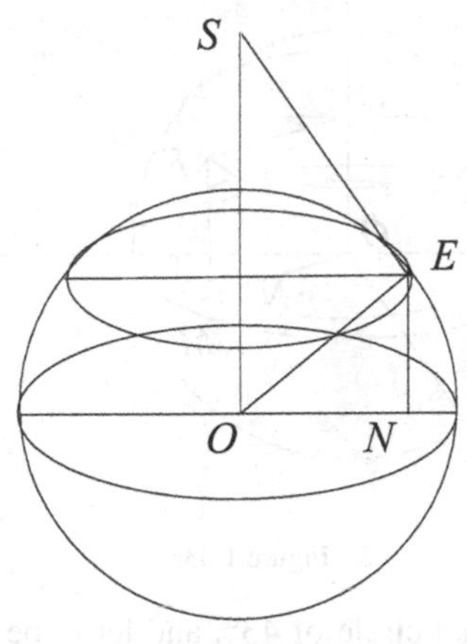

Figure 1.52.

Let N be the point on the equatorial plane that is closest to E. The central angle EON is called the **latitude** of E. If E is in the northern hemisphere, then this angle is positive; otherwise it is negative. Hence the range of latitudes is $[-90°, 90°]$, with $90°$ (or $90°$ north) corresponding to the North Pole and $-90°$ (or $90°$ south) corresponding to the South Pole. Points of constant latitude $x°$ form a circle parallel to the equator. Such a circle is called the **latitudinal circle** of $x°$.

To solve our problem, note that $\cos \angle SOE = \frac{|OE|}{|OS|} = \frac{3960}{4060}$, and so $\angle SOE \approx 12.743°$, implying that $\angle NOE = 90° - \angle SOE \approx 77.254°$. Therefore, all the points with latitude greater than or equal to $77.254°$ can be seen from the space station. ■

Example 1.20. The latitude of the town of Exeter, New Hampshire, is about 43 degrees north.

(a) How far from the equatorial plane is Exeter, assuming that one travels through Earth's interior? What if one travels on Earth's surface?

(b) How far does Earth's rotation carry the citizens of Exeter during a single day?

Solution: Let N be the foot of the perpendicular line segment from E to the equatorial plane. Then $\angle EON = 43°$. In right triangle EON, $|EN| = |EO| \sin 43° = 3960 \sin 43° \approx 2700.714$; that is, the z coordinate of Exeter is about 2700.714 miles. If one travels on Earth's surface, we extend segment ON through N to meet the equator at M. Then $|\widehat{EM}| = \frac{\angle EON}{360°} \cdot 2\pi \cdot 3960 \approx 2971.947$; that is, exeter is about 2971.947 miles away from the equator, assuming that one travels on Earth's surface.

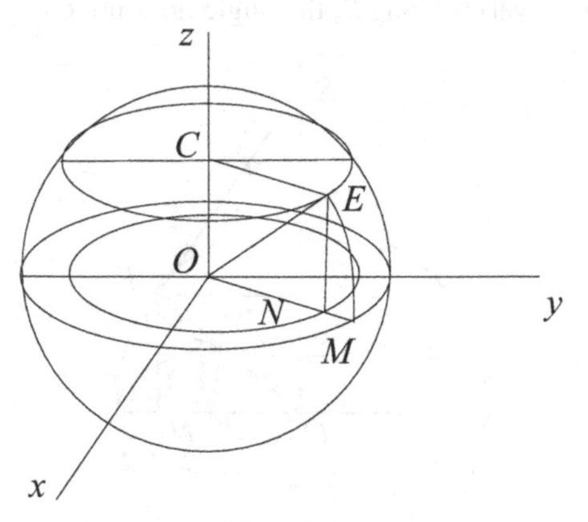

Figure 1.53.

Let $\mathcal{C}$ denote the latitudinal circle of 43°, and let C be the center of $\mathcal{C}$. During a single day (and night; that is, a complete day of 24 hours), Earth's rotation carries the citizens of Exeter through one revolution along the circle $\mathcal{C}$. (If we fix Earth, then E travels along $\mathcal{C}$ for a complete revolution.) Hence, the distance sought is $2\pi \cdot |CE| = 2\pi \cdot |ON| = 7920\pi \cos 43° \approx 18197.114$ miles. (What a long free ride!) ■

It is not difficult to see that points E and N have the same x and y coordinates, and that N lies on a circle centered at O with radius $3960 \cos 43°$. Now we set up the x and y axes. The **primary meridian** is the great semicircle that passes through Greenwich, England, on its way from the North Pole to the South Pole. The x axis is set in such a way that the intersection of the primary meridian and the equator is at (3960, 0, 0). This point is in the South Atlantic Ocean, near the coast of Ghana. The y axis is set such that the positive x, y, and z axes follow the right-hand rule. When we turn the primary meridian around the z axis, we obtain all the semicircles with the segment connecting the North Pole and the South Pole as their common diameter.

These semicircles are called **meridians** or **longitude lines**. We say **longitude** x° (or **longitude line of** x°) if the standard angle of the intersection of the longitude line and the equator is equal to x°, and all the points on this longitude line have longitude x°. Every point on the surface of Earth is the intersection of a latitudinal circle and a longitude line. Hence we can describe the points on the surface by an ordered pair of angles (α, β), where α and β stand for the longitude and latitude angles, respectively. If we also consider points on any sphere, we can write $E = (r, \alpha, \beta)$, where r is the radius of the sphere. These are the **spherical coordinates** of the point E. For the special case of considering points on Earth, r is equal to 3960 miles.

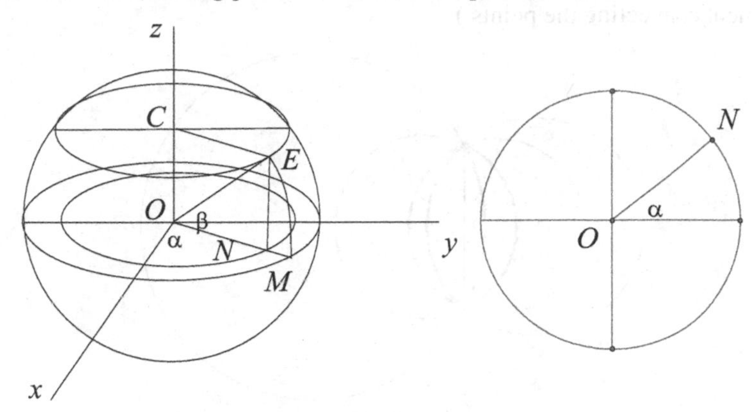

Figure 1.54.

Let E be a point with spherical coordinates (r, α, β) (Figure 1.54). We compute its rectangular coordinates. As we have done before, let N be the foot of the perpendicular line segment from E to the equator. Let $E = (x, y, z)$ be the rectangular coordinates of E. Then $z = |EN| = r \sin\beta$, and $N = (x, y, 0)$. In the xy plane, N lies on a circle with radius $r\cos\beta$, and N has standard angle α. Hence $x = r\cos\alpha\cos\beta$ and $y = r\sin\alpha\cos\beta$; that is,

$$E = (r\cos\alpha\cos\beta, r\sin\alpha\cos\beta, r\sin\beta).$$

Sometimes, we also write $E = r(\cos\alpha\cos\beta, \sin\alpha\cos\beta, \sin\beta)$. This is how to convert spherical coordinates into rectangular coordinates. It is not difficult to reverse this procedure; that is, to convert rectangular coordinates into spherical coordinates.

Traveling on Earth

It is not possible, at least in the foreseeable future, to build tunnels through Earth's interior to connect big cities like New York and Tokyo. Hence, when we travel from place to place on Earth, we need to travel on its surface. For two places on Earth, how can we compute the length of the shortest path along the surface connecting the two places? Let $\mathcal{C}$ denote this path. Intuitively, it is not difficult to see that the points on $\mathcal{C}$ all lie on a plane. This fact is not that easy to prove, however, so let's just accept it. Let $\mathcal{P}$ denote the plane that contains $\mathcal{C}$. It is clear that the intersection of $\mathcal{P}$ and Earth's surface is a circle. Consequently, we conclude that $\mathcal{C}$ is an arc. For two fixed points, we can draw many circles passing through these two points (Figure 1.55). It is not difficult to see that as the radii of the circles increase, the lengths of the minor arcs connecting the points decrease. (When the radius of the circle approaches infinity, the circle becomes a line and the minor arc connecting the points becomes the segment connecting the points.)

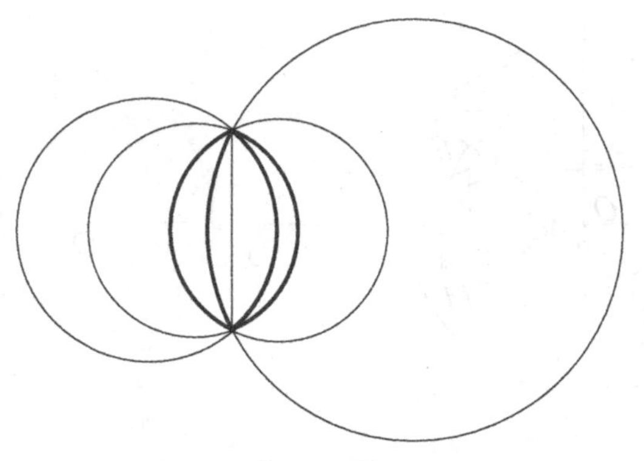

Figure 1.55.

For two points A and B on the surface of Earth, the largest circle passing through A and B on Earth is the circle that is centered at Earth's center. Such circles are called **great circles**. For example, the equator is a great circle, and all points with fixed longitudes form great semicircles. We encourage the reader to choose pairs of arbitrary points on a globe and draw great circles passing through the chosen points. The reader might then have a better idea why many intercontinental airplanes fly at high latitudes.

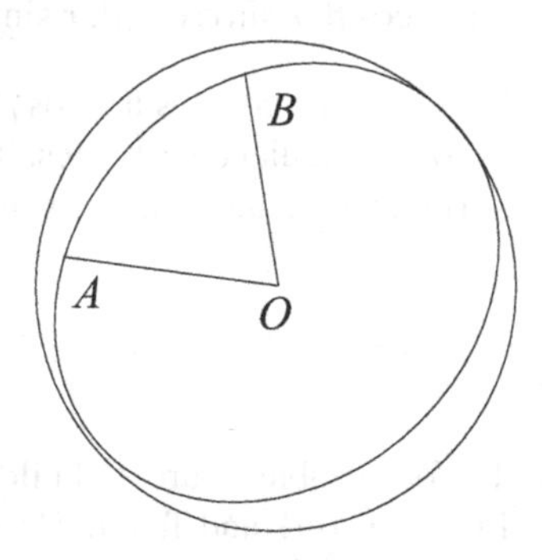

Figure 1.56.

Example 1.21. Monica and Linda traveled from Shanghai, China (east 121°, north 31°), to Albany, New York (west 73°, north 42°), to visit their friend Hilary. Estimate the distance of Monica and Linda's trip.

Solution: In general, let $A = (3960, \alpha_1, \beta_1)$ and $B = (3960, \alpha_2, \beta_2)$. Then

$$A = 3960(\cos\alpha_1 \cos\beta_1, \sin\alpha_1 \cos\beta_1, \sin\beta_1)$$

and

$$B = 3960(\cos\alpha_2 \cos\beta_2, \sin\alpha_2 \cos\beta_2, \sin\beta_2).$$

Let $\theta = \angle AOB$. Then by the vector form of the **law of cosines**, we have

$$\begin{aligned}\cos\theta &= \frac{\overrightarrow{OA}\cdot\overrightarrow{OB}}{|\overrightarrow{OA}||\overrightarrow{OB}|}\\ &= \cos\alpha_1 \cos\beta_1 \cos\alpha_2 \cos\beta_2 + \sin\alpha_1 \cos\beta_1 \sin\alpha_2 \cos\beta_2 + \sin\beta_1 \sin\beta_2.\end{aligned}$$

For this problem, we have $\alpha_1 = 121°$, $\beta_1 = 31°$, $\alpha_2 = -73°$, and $\beta_2 = 42°$. Plugging these values into the above equation gives $\cos\theta \approx -0.273$, implying that $\theta \approx 105.870°$. Hence the distance between Shanghai and Albany is about $\frac{\theta}{360°} \cdot 2\pi \cdot 3960 \approx 7317.786$ miles along Earth's surface.

Where Are You?

During a long flight, the screens in the plane cabin often show the position of the plane and the trajectory of the trip. How is it done? Or, how does a GPS (Global Positioning System) work? Let's give a baby tour of this subject.

Example 1.22. Monica and Linda traveled from Shanghai, China (east 121°, north 31°), to Albany, New York (west 73°, north 42°), to visit their friend Hilary. Monica's hometown, Billrock, is four-fifths of the way along their trip. Find the spherical coordinates of the town of Billrock.

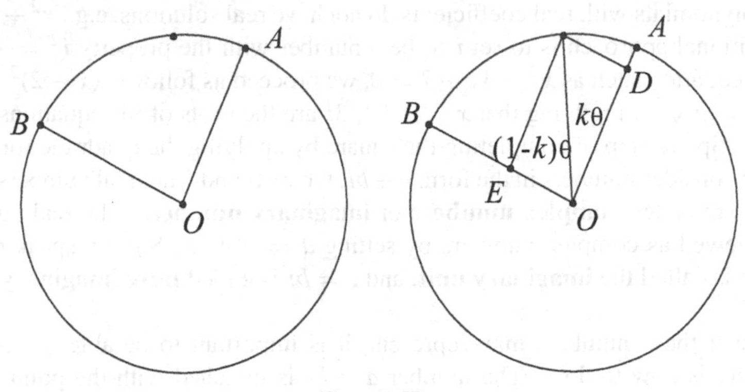

Figure 1.57.

Solution: We maintain the same notation as in the solution of Example 1.21. Let C denote an arbitrary point on $\widehat{AB}$, and assume that $\angle AOC = k\angle AOB = k\theta$, where is k is some real number with $0 \le k \le 1$. (See Figure 1.57.) Then $\angle COB = (1-k)\theta$. Let D and E be points on lines OA and OB, respectively, such that $AO \parallel CE$ and $BO \parallel CD$. Then $CDOE$ is parallelogram. Set $\mathbf{u} = \overrightarrow{OA}$, $\mathbf{v} = \overrightarrow{OB}$, and $\mathbf{w} = \overrightarrow{OC}$. There are real numbers a and b such that $|OD| = a \cdot |OA|$ and $|OE| = b \cdot |OB|$. Then $\mathbf{w} = a\mathbf{u} + b\mathbf{v}$. By the distributive property of the dot product, we have

$$\mathbf{w} \cdot \mathbf{u} = a\mathbf{u} \cdot \mathbf{u} + b\mathbf{u} \cdot \mathbf{v} \quad \text{and} \quad \mathbf{w} \cdot \mathbf{v} = a\mathbf{u} \cdot \mathbf{v} + b\mathbf{v} \cdot \mathbf{v}.$$

Note that $|\mathbf{u}| = |\mathbf{v}| = |\mathbf{w}| = 3960$. By the vector form of the **law of cosines**, these equations are equivalent to the equations

$$\cos k\theta = a + b\cos\theta \quad \text{and} \quad \cos(1-k)\theta = a\cos\theta + b.$$

Solving this system of equations for a and b gives

$$a = \frac{\cos k\theta - \cos(1-k)\theta\cos\theta}{\sin^2\theta} \quad \text{and} \quad b = \frac{\cos(1-k)\theta - \cos k\theta\cos\theta}{\sin^2\theta}.$$

For our problem, we have $k = \frac{4}{5}$ and $\theta \approx 105.870°$ (by Example 1.21). Substituting these values into the above equations gives $a \approx 0.376$ and $b \approx 1.035$. It follows that $\mathbf{w} = a\mathbf{u} + b\mathbf{v} \approx 3960[0.059, -0.460, 0.886]$, which are the rectangular coordinates of the point C. Let $(3960, \gamma, \phi)$ be the spherical coordinates of C. Then $\sin\phi \approx 0.886$, or $\phi \approx 62.383°$, and $\sin\gamma\cos\phi \approx -0.460$, or $\gamma \approx -82.67°$; that is, Billrock has longitude west 82.67° and latitude 62.383°.

De Moivre's Formula

Many polynomials with real coefficients do not have real solutions, e.g., $x^2 + 1 = 0$. The traditional approach is to set i to be a number with the property $i^2 = -1$. To solve an equation such as $x^2 - 4x + 7 = 0$, we proceed as follows: $(x-2)^2 = -3$, or $x - 2 = \pm\sqrt{3}i$, implying that $x = 2 \pm \sqrt{3}i$ are the roots of the equations. (One can also skip the step of completing the square by applying the quadratic formula.) Thus, we consider numbers in the form $a + bi$, where a and b are real numbers. Such numbers are called **complex numbers** or **imaginary numbers**. The real numbers can be viewed as complex numbers by setting $a = a + 0i$. Strictly speaking, the number i is called the **imaginary unit**, and $z = bi$ is called **pure imaginary** if b is nonzero.

Whatever these numbers may represent, it is important to be able to visualize them. Here is how to do it: The number $a + bi$ is matched with the point (a, b), or the vector $[a, b]$ that points from the origin to (a, b). Under this convention, the coordinate plane is called the **complex plane**. Points $(0, b)$ on the y axis are thereby matched with pure imaginary numbers bi, so the y axis is called the **imaginary axis** in the complex plane. Similarly, the x axis is called the **real axis**. Let O denote the origin, and let each lowercase letter denote the complex number assigned to the point labeled with the corresponding uppercase letter. (For example, if $z = 3 + 4i$, then $Z = (3, 4)$.) See Figure 1.58.

The definition of the complex plane allows us to talk about operations on complex numbers. The sum of the complex numbers $z_1 = a_1 + b_1 i$ and $z_2 = a_2 + b_2 i$ is $z = z_1 + z_2 = (a_1 + a_2) + (b_1 + b_2)i$, and their difference is $z' = z_1 - z_2 = (a_1 - a_2) + (b_1 - b_2)i$. Because of the vector interpretation of complex numbers, it is not difficult to see that OZ_1ZZ_2 forms a parallelogram, and z and z' are matched with diagonal vectors $\overrightarrow{OZ}$ and $\overrightarrow{Z_2Z_1}$.

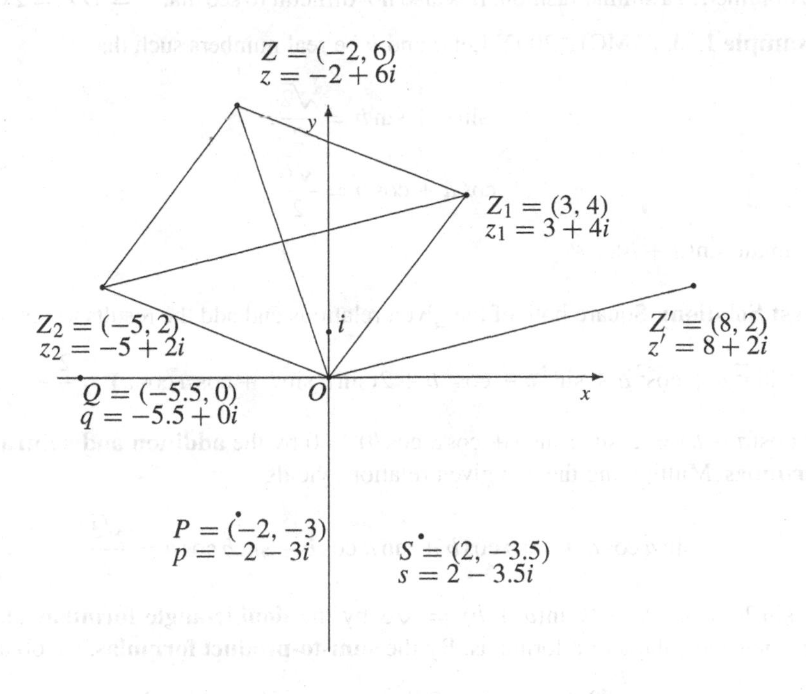

Figure 1.58.

We can talk about the **magnitude** or **length** of a complex number z, denoted by $|z|$, by considering the magnitude of the vector it corresponds to. For example, $|3-4i| = 5$, and in general, $|a+bi| = \sqrt{a^2 + b^2}$. Hence, all the complex numbers z with $|z| = 5$ form a circle centered at the origin with radius 5. We can also talk about the **polar form** of complex numbers. If $Z = (a, b)$ has polar coordinates $Z = (r, \theta)$, then $r = \sqrt{a^2 + b^2}$ and $a = r\cos\theta$ and $b = r\sin\theta$. We write $z = a + bi = r\cos\theta + ri\sin\theta = r\operatorname{cis}\theta$. For example, $z = -1 + \sqrt{3}i = 2\operatorname{cis}120^\circ = 2\operatorname{cis}\frac{2\pi}{3}$.

We have developed interesting properties of adding vectors with equal magnitudes. Similar properties can be written in terms of complex numbers. Let z_1 and z_2 be complex numbers with $|z_1| = |z_2| = r$. Set $z_1 = r\operatorname{cis}\theta_1$ and $z_2 = r\operatorname{cis}\theta_2$. Set $z = z_1 + z_2$. Then OZ_1ZZ_2 is a rhombus, implying that the line OZ bisects the angle Z_1OZ_2; that is, $z = r'\operatorname{cis}\frac{\theta_1+\theta_2}{2}$ for some real number r'. In particular, if $z_1 = 1$ and $z_2 = \operatorname{cis}a$, then $z = r'\operatorname{cis}\frac{a}{2}$, where $r' = \overline{OZ}$. Therefore, $\tan\frac{a}{2}$ is equal to the slope of line OZ; that is,

$$\tan\frac{a}{2} = \frac{\sin a}{1 + \cos a},$$

which is one of **half-angle formulas**. Other versions of the half-angle formulas can be obtained in a similar fashion. It is also not difficult to see that $r' = \overline{OZ} = 2\cos\frac{a}{2}$.

Example 1.23. [AMC12 2002] Let a and b be real numbers such that

$$\sin a + \sin b = \frac{\sqrt{2}}{2},$$
$$\cos a + \cos b = \frac{\sqrt{6}}{2}.$$

Evaluate $\sin(a + b)$.

First Solution: Square both of the given relations and add the results to obtain

$$\sin^2 a + \cos^2 a + \sin^2 b + \cos^2 b + 2(\sin a \sin b + \cos a \cos b) = \frac{2}{4} + \frac{6}{4},$$

or $\cos(a - b) = 2(\sin a \sin b + \cos a \cos b) = 0$ by the **addition and subtraction formulas**. Multiplying the two given relations yields

$$\sin a \cos a + \sin b \cos b + \sin a \cos b + \sin b \cos a = \frac{\sqrt{3}}{2},$$

or $\sin 2a + \sin 2b + 2\sin(a + b) = \sqrt{3}$ by the **double-angle formulas** and the addition and subtraction formulas. By the **sum-to-product formulas**, we obtain

$$\sin 2a + \sin 2b = 2\sin(a + b)\cos(a - b) = 0.$$

Hence $\sin(a + b) = \frac{\sqrt{3}}{2}$.

Second Solution:

Set complex numbers $z_1 = \cos a + i\sin a = \operatorname{cis}a$ and $z_2 = \cos b + i\sin b = \operatorname{cis}b$ (Figure 1.59).

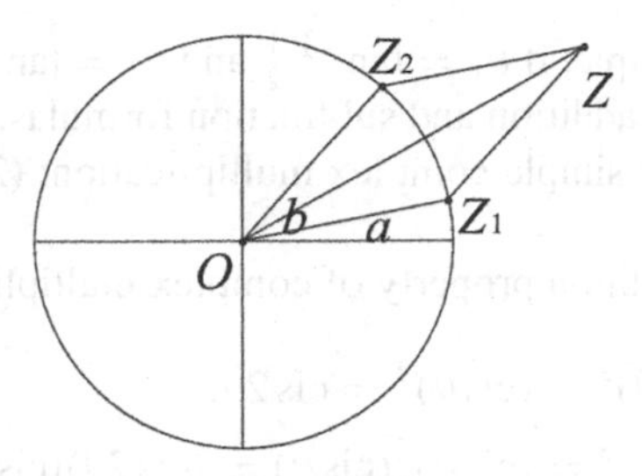

Figure 1.59.

Then by the **operation rules for complex numbers** and by the given conditions, we have

$$\begin{aligned}z = z_1 + z_2 = r \operatorname{cis} \frac{a+b}{2} &= \operatorname{cis} a + \operatorname{cis} b \\ = \frac{\sqrt{6}}{2} + \frac{\sqrt{2}i}{2} &= \sqrt{2}\left(\frac{\sqrt{3}}{2} + \frac{i}{2}\right) = \sqrt{2} \operatorname{cis} \frac{\pi}{6};\end{aligned}$$

that is, $r' = \sqrt{2}$ and $\frac{a+b}{2} = \frac{\pi}{6}$, and so $\sin(a+b) = \sin\frac{\pi}{3} = \frac{\sqrt{3}}{2}$. ■

One of the most interesting properties of complex numbers is related to the product. It is natural to define the product of complex numbers $z_1 = a_1 + b_1 i$ and $z_2 = a_2 + b_2 i$ as

$$\begin{aligned}z = z_1 z_2 &= (a_1 + b_1 i)(a_2 + b_2 i) \\ &= a_1 a_2 + a_2 b_1 i + a_1 b_2 i + b_1 b_2 i^2 \\ &= (a_1 a_2 - b_1 b_2) + (a_1 b_2 + a_2 b_1) i.\end{aligned}$$

This certainly mimics the form of the **addition and subtraction formulas**. Indeed, working in the polar form $z_1 = r_1 \operatorname{cis} \theta_1$ and $z_2 = r_2 \operatorname{cis} \theta_2$, we have

$$\begin{aligned}z = z_1 z_2 &= (a_1 + b_1 i)(a_2 + b_2 i) = (r_1 \operatorname{cis} \theta_1)(r_2 \operatorname{cis} \theta_2) \\ &= r_1(\cos\theta_1 + \sin\theta_1 i) r_2(\cos\theta_2 + \sin\theta_2 i) \\ &= r_1 r_2(\cos\theta_1 + \sin\theta_1 i)(\cos\theta_2 + \sin\theta_2 i) \\ &= r_1 r_2[\cos(\theta_1 + \theta_2) + i\sin(\theta_1 + \theta_2)] \\ &= r_1 r_2 \operatorname{cis}(\theta_1 + \theta_2),\end{aligned}$$

by noting the addition and subtraction formulas $\cos(\theta_1 + \theta_2) = \cos\theta_1\cos\theta_2 - \sin\theta_1\sin\theta_2$ and $\sin(\theta_1 + \theta_2) = \sin\theta_1\cos\theta_2 + \cos\theta_1\sin\theta_2$. It is then not difficult to see that

$$\frac{z_1}{z_2} = \frac{r_1 \operatorname{cis} \theta_1}{r_2 \operatorname{cis} \theta_2} = \frac{r_1}{r_2} \operatorname{cis}(\theta_1 - \theta_2).$$

We have proved the angle-addition (angle-subtraction) property of complex multiplication (division). This is why operations on complex numbers are closely related to trigonometry. For example, if $\theta_1 = \tan^{-1}\frac{1}{2}$ and $\theta_2 = \tan^{-1}\frac{1}{3}$, then we can show that $\theta_1 + \theta_2 = 45^\circ$ by the addition and subtraction formulas. On the other hand, this fact can also be shown by simple complex multiplication: $(2+i)(3+i) = 5+5i$. Can the reader tell why?

By the above angle-addition property of complex multiplication, we can see that

$$\begin{aligned}(\cos\theta + i\sin\theta)^2 &= (\operatorname{cis}\theta)^2 = \operatorname{cis} 2\theta,\\ (\cos\theta + i\sin\theta)^3 &= (\operatorname{cis}\theta)^2(\operatorname{cis}\theta) = (\operatorname{cis} 2\theta)(\operatorname{cis}\theta) = \operatorname{cis} 3\theta,\end{aligned}$$

and so on. By this induction process, we can prove **de Moivre's formula**: For any angle θ and for any integer n,

$$(\cos\theta + i\sin\theta)^n = (\operatorname{cis}\theta)^n = \operatorname{cis} n\theta = \cos n\theta + i\sin n\theta.$$

From this formula, it is not difficult to derive the **expansion formulas** of $\sin n\alpha$ and $\cos n\alpha$ in terms of $\sin\alpha$ and $\cos\alpha$ by expanding the left-hand side of the above identity and matching the corresponding real and imaginary parts of both sides:

$$\begin{aligned}\sin n\alpha &= \binom{n}{1}\cos^{n-1}\alpha\sin\alpha - \binom{n}{3}\cos^{n-3}\alpha\sin^3\alpha\\ &\quad + \binom{n}{5}\cos^{n-5}\alpha\sin^5\alpha - \cdots,\\ \cos n\alpha &= \binom{n}{0}\cos^{n}\alpha - \binom{n}{2}\cos^{n-2}\alpha\sin^2\alpha\\ &\quad + \binom{n}{4}\cos^{n-4}\alpha\sin^4\alpha - \cdots.\end{aligned}$$

2

Introductory Problems

1. Let x be a real number such that $\sec x - \tan x = 2$. Evaluate $\sec x + \tan x$.

2. Let $0^\circ < \theta < 45^\circ$. Arrange

$$t_1 = (\tan\theta)^{\tan\theta}, \qquad t_2 = (\tan\theta)^{\cot\theta},$$
$$t_3 = (\cot\theta)^{\tan\theta}, \qquad t_4 = (\cot\theta)^{\cot\theta},$$

in decreasing order.

3. Compute

   (a) $\sin\frac{\pi}{12}$, $\cos\frac{\pi}{12}$, and $\tan\frac{\pi}{12}$;
   (b) $\cos^4\frac{\pi}{24} - \sin^4\frac{\pi}{24}$;
   (c) $\cos 36^\circ - \cos 72^\circ$; and
   (d) $\sin 10^\circ \sin 50^\circ \sin 70^\circ$.

4. Simplify the expression

$$\sqrt{\sin^4 x + 4\cos^2 x} - \sqrt{\cos^4 x + 4\sin^2 x}.$$

5. Prove that

$$1 - \cot 23^\circ = \frac{2}{1 - \cot 22^\circ}.$$

6. Find all x in the interval $\left(0, \frac{\pi}{2}\right)$ such that

$$\frac{\sqrt{3} - 1}{\sin x} + \frac{\sqrt{3} + 1}{\cos x} = 4\sqrt{2}.$$

7. Region $\mathcal{R}$ contains all the points (x, y) such that $x^2 + y^2 \le 100$ and $\sin(x + y) \ge 0$. Find the area of region $\mathcal{R}$.

8. In triangle ABC, show that

$$\sin \frac{A}{2} \le \frac{a}{b + c}.$$

9. Let I denote the interval $\left[-\frac{\pi}{4}, \frac{\pi}{4}\right]$. Determine the function f defined on the interval $[-1, 1]$ such that $f(\sin 2x) = \sin x + \cos x$ and simplify $f(\tan^2 x)$ for x in the interval I.

10. Let

$$f_k(x) = \frac{1}{k}(\sin^k x + \cos^k x)$$

for $k = 1, 2, \ldots$. Prove that

$$f_4(x) - f_6(x) = \frac{1}{12}$$

for all real numbers x.

11. A circle of radius 1 is randomly placed in a 15-by-36 rectangle $ABCD$ so that the circle lies completely within the rectangle. Given that the probability that the circle will not touch diagonal AC is $\frac{m}{n}$, where m and n are relatively prime positive integers, find $m + n$.

12. In triangle ABC,

$$3\sin A + 4\cos B = 6 \quad \text{and} \quad 4\sin B + 3\cos A = 1.$$

Find the measure of angle C.

13. Prove that

$$\tan 3a - \tan 2a - \tan a = \tan 3a \tan 2a \tan a$$

for all $a \neq \frac{k\pi}{2}$, where k is in $\mathbb{Z}$.

14. Let a, b, c, d be numbers in the interval $[0, \pi]$ such that

$$\sin a + 7\sin b = 4(\sin c + 2\sin d),$$
$$\cos a + 7\cos b = 4(\cos c + 2\cos d).$$

Prove that $2\cos(a - d) = 7\cos(b - c)$.

15. Express

$$\sin(x - y) + \sin(y - z) + \sin(z - x)$$

as a monomial.

16. Prove that

$$(4\cos^2 9^\circ - 3)(4\cos^2 27^\circ - 3) = \tan 9^\circ.$$

17. Prove that

$$\left(1 + \frac{a}{\sin x}\right)\left(1 + \frac{b}{\cos x}\right) \geq \left(1 + \sqrt{2ab}\right)^2$$

for all real numbers a, b, x with $a, b \geq 0$ and $0 < x < \frac{\pi}{2}$.

18. In triangle ABC, $\sin A + \sin B + \sin C \leq 1$. Prove that

$$\min\{A + B, B + C, C + A\} < 30^\circ.$$

19. Let ABC be a triangle. Prove that

(a)

$$\tan\frac{A}{2}\tan\frac{B}{2}+\tan\frac{B}{2}\tan\frac{C}{2}+\tan\frac{C}{2}\tan\frac{A}{2}=1;$$

(b)

$$\tan\frac{A}{2}\tan\frac{B}{2}\tan\frac{C}{2}\le\frac{\sqrt{3}}{9}.$$

20. Let ABC be an acute-angled triangle. Prove that

(a) $\tan A+\tan B+\tan C=\tan A\tan B\tan C$;

(b) $\tan A\tan B\tan C\ge 3\sqrt{3}$.

21. Let ABC be a triangle. Prove that

$$\cot A\cot B+\cot B\cot C+\cot C\cot A=1.$$

Conversely, prove that if x, y, z are real numbers with $xy+yz+zx=1$, then there exists a triangle ABC such that $\cot A=x$, $\cot B=y$, and $\cot C=z$.

22. Let ABC be a triangle. Prove that

$$\sin^2\frac{A}{2}+\sin^2\frac{B}{2}+\sin^2\frac{C}{2}+2\sin\frac{A}{2}\sin\frac{B}{2}\sin\frac{C}{2}=1.$$

Conversely, prove that if x, y, z are positive real numbers such that

$$x^2+y^2+z^2+2xyz=1,$$

then there is a triangle ABC such that $x=\sin\frac{A}{2}$, $y=\sin\frac{B}{2}$, and $z=\sin\frac{C}{2}$.

23. Let ABC be a triangle. Prove that

(a) $\sin\frac{A}{2}\sin\frac{B}{2}\sin\frac{C}{2}\le\frac{1}{8}$;

(b) $\sin^2\frac{A}{2}+\sin^2\frac{B}{2}+\sin^2\frac{C}{2}\ge\frac{3}{4}$;

(c) $\cos^2\frac{A}{2}+\cos^2\frac{B}{2}+\cos^2\frac{C}{2}\le\frac{9}{4}$;

(d) $\cos\frac{A}{2}\cos\frac{B}{2}\cos\frac{C}{2}\le\frac{3\sqrt{3}}{8}$;

(e) $\csc\frac{A}{2}+\csc\frac{A}{2}+\csc\frac{A}{2}\geq 6.$

24. In triangle ABC, show that

(a) $\sin 2A+\sin 2B+\sin 2C=4\sin A\sin B\sin C$;

(b) $\cos 2A+\cos 2B+\cos 2C=-1-4\cos A\cos B\cos C$;

(c) $\sin^2 A+\sin^2 B+\sin^2 C=2+2\cos A\cos B\cos C$;

(d) $\cos^2 A+\cos^2 B+\cos^2 C+2\cos A\cos B\cos C=1$.

Conversely, if x, y, z are positive real numbers such that

$$x^2+y^2+z^2+2xyz=1,$$

show that there is an acute triangle ABC such that $x=\cos A$, $y=\cos B$, $C=\cos C$.

25. In triangle ABC, show that

(a) $4R=\frac{abc}{[ABC]}$;

(b) $2R^2\sin A\sin B\sin C=[ABC]$;

(c) $2R\sin A\sin B\sin C=r(\sin A+\sin B+\sin C)$;

(d) $r=4R\sin\frac{A}{2}\sin\frac{B}{2}\sin\frac{C}{2}$;

(e) $a\cos A+b\cos B+c\cos C=\frac{abc}{2R^2}$.

26. Let s be the semiperimeter of triangle ABC. Prove that

(a) $s=4R\cos\frac{A}{2}\cos\frac{B}{2}\cos\frac{C}{2}$;

(b) $s\leq\frac{3\sqrt{3}}{2}R$.

27. In triangle ABC, show that

(a) $\cos A+\cos B+\cos C=1+4\sin\frac{A}{2}\sin\frac{B}{2}\sin\frac{C}{2}$;

(b) $\cos A + \cos B + \cos C \le \frac{3}{2}$.

28. Let ABC be a triangle. Prove that

(a) $\cos A \cos B \cos C \le \frac{1}{8}$;

(b) $\sin A \sin B \sin C \le \frac{3\sqrt{3}}{8}$;

(c) $\sin A + \sin B + \sin C \le \frac{3\sqrt{3}}{2}$.

(d) $\cos^2 A + \cos^2 B + \cos^2 C \ge \frac{3}{4}$;

(e) $\sin^2 A + \sin^2 B + \sin^2 C \le \frac{9}{4}$;

(f) $\cos 2A + \cos 2B + \cos 2C \ge -\frac{3}{2}$;

(g) $\sin 2A + \sin 2B + \sin 2C \le \frac{3\sqrt{3}}{2}$.

29. Prove that

$$\frac{\tan 3x}{\tan x} = \tan\left(\frac{\pi}{3} - x\right)\tan\left(\frac{\pi}{3} + x\right)$$

for all $x \neq \frac{k\pi}{6}$, where k is in $\mathbb{Z}$.

30. Given that

$$(1 + \tan 1^\circ)(1 + \tan 2^\circ) \cdots (1 + \tan 45^\circ) = 2^n,$$

find n.

31. Let $A = (0, 0)$ and $B = (b, 2)$ be points in the coordinate plane. Let $ABCDEF$ be a convex equilateral hexagon such that $\angle FAB = 120^\circ$, $AB \parallel DE$, $BC \parallel EF$, and $CD \parallel FA$, and the y coordinates of its vertices are distinct elements of the set $\{0, 2, 4, 6, 8, 10\}$. The area of the hexagon can be written in the form $m\sqrt{n}$, where m and n are positive integers and n is not divisible by the square of any prime. Find $m + n$.

32. Show that one can use a composition of trigonometry buttons, such as sin, cos, tan, $\sin^{-1}$, $\cos^{-1}$, and $\tan^{-1}$, to replace the broken reciprocal button on a calculator.

33. In triangle ABC, $A - B = 120°$ and $R = 8r$. Find $\cos C$.

34. Prove that in a triangle ABC,

$$\frac{a-b}{a+b} = \tan\frac{A-B}{2}\tan\frac{C}{2}.$$

35. In triangle ABC, $\frac{a}{b} = 2 + \sqrt{3}$ and $\angle C = 60°$. Find the measure of angles A and B.

36. Let a, b, c be real numbers, all different from -1 and 1, such that $a+b+c = abc$. Prove that

$$\frac{a}{1-a^2} + \frac{b}{1-b^2} + \frac{c}{1-c^2} = \frac{4abc}{(1-a^2)(1-b^2)(1-c^2)}.$$

37. Prove that a triangle ABC is isosceles if and only if

$$a\cos B + b\cos C + c\cos A = \frac{a+b+c}{2}.$$

38. Evaluate

$$\cos a \cos 2a \cos 3a \cdots \cos 999a,$$

where $a = \frac{2\pi}{1999}$.

39. Determine the minimum value of

$$\frac{\sec^4\alpha}{\tan^2\beta} + \frac{\sec^4\beta}{\tan^2\alpha}$$

over all $\alpha, \beta \neq \frac{k\pi}{2}$, where k is in $\mathbb{Z}$.

40. Find all pairs (x, y) of real numbers with $0 < x < \frac{\pi}{2}$ such that

$$\frac{(\sin x)^{2y}}{(\cos x)^{y^2/2}} + \frac{(\cos x)^{2y}}{(\sin x)^{y^2/2}} = \sin 2x.$$

41. Prove that $\cos 1°$ is an irrational number.

42. Find the maximum value of

$$S = (1 - x_1)(1 - y_1) + (1 - x_2)(1 - y_2)$$

if $x_1^2 + x_2^2 = y_1^2 + y_2^2 = c^2$.

43. Prove that

$$\frac{\sin^3 a}{\sin b} + \frac{\cos^3 a}{\cos b} \ge \sec(a - b)$$

for all $0 < a, b < \frac{\pi}{2}$.

44. If $\sin\alpha\cos\beta = -\frac{1}{2}$, what are the possible values of $\cos\alpha\sin\beta$?

45. Let a, b, c be real numbers. Prove that

$$(ab + bc + ca - 1)^2 \le (a^2 + 1)(b^2 + 1)(c^2 + 1).$$

46. Prove that

$$(\sin x + a\cos x)(\sin x + b\cos x) \le 1 + \left(\frac{a+b}{2}\right)^2.$$

47. Prove that

$$|\sin a_1| + |\sin a_2| + \cdots + |\sin a_n| + |\cos(a_1 + a_2 + \cdots + a_n)| \ge 1.$$

48. Find all angles α for which the three-element set

$$S = \{\sin\alpha, \sin 2\alpha, \sin 3\alpha\}$$

is equal to the set

$$T = \{\cos\alpha, \cos 2\alpha, \cos 3\alpha\}.$$

49. Let $\{T_n(x)\}_{n=0}^{\infty}$ be the sequence of polynomials such that $T_0(x) = 1$, $T_1(x) = x$, $T_{i+1} = 2xT_i(x) - T_{i-1}(x)$ for all positive integers i. The polynomial $T_n(x)$ is called the nth **Chebyshev polynomial**.

(a) Prove that $T_{2n+1}(x)$ and $T_{2n}(x)$ are odd and even functions, respectively;

(b) Prove that $T_n(\cos\theta) = \cos(n\theta)$ for all nonnegative integers n;

(c) Prove that for real numbers x in the interval $[-1, 1]$, $-1 \le T_n(x) \le 1$;

(d) Prove that $T_{n+1}(x) > T_n(x)$ for real numbers x with $x > 1$;

(e) Determine all the roots of $T_n(x)$;

(f) Determine all the roots of $P_n(x) = T_n(x) - 1$.

50. Let ABC be a triangle with $\angle BAC = 40°$ and $\angle ABC = 60°$. Let D and E be the points lying on the sides AC and AB, respectively, such that $\angle CBD = 40°$ and $\angle BCE = 70°$. Segments BD and CE meet at F. Show that $AF \perp BC$.

51. Let S be an interior point of triangle ABC. Show that at least one of $\angle SAB$, $\angle SBC$, and $\angle SCA$ is less than or equal to $30°$.

52. Let $a = \frac{\pi}{7}$.

(a) Show that $\sin^2 3a - \sin^2 a = \sin 2a \sin 3a$;

(b) Show that $\csc a = \csc 2a + \csc 4a$;

(c) Evaluate $\cos a - \cos 2a + \cos 3a$;

(d) Prove that $\cos a$ is a root of the equation $8x^3 + 4x^2 - 4x - 1 = 0$;

(e) Prove that $\cos a$ is irrational;

(f) Evaluate $\tan a \tan 2a \tan 3a$;

(g) Evaluate $\tan^2 a + \tan^2 2a + \tan^2 3a$;

(h) Evaluate $\tan^2 a \tan^2 2a + \tan^2 2a \tan^2 3a + \tan^2 3a \tan^2 a$.

(i) Evaluate $\cot^2 a + \cot^2 2a + \cot^2 3a$.

3
Advanced Problems

1. Two exercises on $\sin k^\circ \sin(k+1)^\circ$:

   (a) Find the smallest positive integer n such that

$$\frac{1}{\sin 45^\circ \sin 46^\circ} + \frac{1}{\sin 47^\circ \sin 48^\circ} + \cdots + \frac{1}{\sin 133^\circ \sin 134^\circ} = \frac{1}{\sin n^\circ}.$$

   (b) Prove that

$$\frac{1}{\sin 1^\circ \sin 2^\circ} + \frac{1}{\sin 2^\circ \sin 3^\circ} + \cdots + \frac{1}{\sin 89^\circ \sin 90^\circ} = \frac{\cos 1^\circ}{\sin^2 1^\circ}.$$

2. Let ABC be a triangle, and let x be a nonnegative real number. Prove that

$$a^x \cos A + b^x \cos B + c^x \cos C \le \frac{1}{2}(a^x + b^x + c^x).$$

3. Let x, y, z be positive real numbers.

(a) Prove that

$$\frac{x}{\sqrt{1+x^2}}+\frac{y}{\sqrt{1+y^2}}+\frac{z}{\sqrt{1+z^2}}\le\frac{3\sqrt{3}}{2}$$

if $x+y+z=xyz$;

(b) Prove that

$$\frac{x}{1-x^2}+\frac{y}{1-y^2}+\frac{z}{1-z^2}\ge\frac{3\sqrt{3}}{2}$$

if $0<x,y,z<1$ and $xy+yz+zx=1$.

4. Let x,y,z be real numbers with $x\ge y\ge z\ge\frac{\pi}{12}$ such that $x+y+z=\frac{\pi}{2}$. Find the maximum and the minimum values of the product $\cos x\sin y\cos z$.

5. Let ABC be an acute-angled triangle, and for $n=1,2,3$, let

$$x_n=2^{n-3}(\cos^n A+\cos^n B+\cos^n C)+\cos A\cos B\cos C.$$

Prove that

$$x_1+x_2+x_3\ge\frac{3}{2}.$$

6. Find the sum of all x in the interval $[0,2\pi]$ such that

$$3\cot^2 x+8\cot x+3=0.$$

7. Let ABC be an acute-angled triangle with side lengths a,b,c and area K. Prove that

$$\sqrt{a^2b^2-4K^2}+\sqrt{b^2c^2-4K^2}+\sqrt{c^2a^2-4K^2}=\frac{a^2+b^2+c^2}{2}.$$

8. Compute the sums

$$\binom{n}{1}\sin a+\binom{n}{2}\sin 2a+\cdots+\binom{n}{n}\sin na$$

and

$$\binom{n}{1}\cos a+\binom{n}{2}\cos 2a+\cdots+\binom{n}{n}\cos na.$$

9. Find the minimum value of

$$|\sin x + \cos x + \tan x + \cot x + \sec x + \csc x|$$

for real numbers x.

10. Two real sequences $x_1, x_2, \ldots$ and $y_1, y_2, \ldots$ are defined in the following way:

$$x_1 = y_1 = \sqrt{3}, \quad x_{n+1} = x_n + \sqrt{1 + x_n^2}, \quad y_{n+1} = \frac{y_n}{1 + \sqrt{1 + y_n^2}},$$

for all $n \geq 1$. Prove that $2 < x_n y_n < 3$ for all $n > 1$.

11. Let a, b, c be real numbers such that

$$\sin a + \sin b + \sin c \geq \frac{3}{2}.$$

Prove that

$$\sin\left(a - \frac{\pi}{6}\right) + \sin\left(b - \frac{\pi}{6}\right) + \sin\left(c - \frac{\pi}{6}\right) \geq 0.$$

12. Consider any four numbers in the interval $\left[\frac{\sqrt{2}-\sqrt{6}}{2}, \frac{\sqrt{2}+\sqrt{6}}{2}\right]$. Prove that there are two of them, say a and b, such that

$$\left|a\sqrt{4 - b^2} - b\sqrt{4 - a^2}\right| \leq 2.$$

13. Let a and b be real numbers in the interval $[0, \frac{\pi}{2}]$. Prove that

$$\sin^6 a + 3\sin^2 a \cos^2 b + \cos^6 b = 1$$

if and only if $a = b$.

14. Let x, y, z be real numbers with $0 < x < y < z < \frac{\pi}{2}$. Prove that

$$\frac{\pi}{2} + 2\sin x \cos y + 2\sin y \cos z \geq \sin 2x + \sin 2y + \sin 2z.$$

15. For a triangle XYZ, let r_{XYZ} denote its inradius. Given that the convex pentagon $ABCDE$ is inscribed in a circle, prove that if $r_{ABC} = r_{AED}$ and $r_{ABD} = r_{AEC}$, then triangles ABC and AED are congruent.

16. All the angles in triangle ABC are less then 120°. Prove that

$$\frac{\cos A + \cos B - \cos C}{\sin A + \sin B - \sin C} > -\frac{\sqrt{3}}{3}.$$

17. Let ABC be a triangle such that

$$\left(\cot \frac{A}{2}\right)^2 + \left(2\cot \frac{B}{2}\right)^2 + \left(3\cot \frac{C}{2}\right)^2 = \left(\frac{6s}{7r}\right)^2,$$

where s and r denote its semiperimeter and its inradius, respectively. Prove that triangle ABC is similar to a triangle T whose side lengths are all positive integers with no common divisor and determine these integers.

18. Prove that the average of the numbers

$$2\sin 2°,\quad 4\sin 4°,\quad 6\sin 6°,\quad \ldots,\quad 180\sin 180°$$

is $\cot 1°$.

19. Prove that in any acute triangle ABC,

$$\cot^3 A + \cot^3 B + \cot^3 C + 6\cot A\cot B\cot C \geq \cot A + \cot B + \cot C.$$

20. Let $\{a_n\}$ be the sequence of real numbers defined by $a_1 = t$ and $a_{n+1} = 4a_n(1-a_n)$ for $n \geq 1$. For how many distinct values of t do we have $a_{2004} = 0$?

21. Triangle ABC has the following property: there is an interior point P such that $\angle PAB = 10°$, $\angle PBA = 20°$, $\angle PCA = 30°$, and $\angle PAC = 40°$. Prove that triangle ABC is isosceles.

22. Let $a_0 = \sqrt{2} + \sqrt{3} + \sqrt{6}$, and let $a_{n+1} = \frac{a_n^2 - 5}{2(a_n + 2)}$ for integers $n > 0$. Prove that

$$a_n = \cot\left(\frac{2^{n-3}\pi}{3}\right) - 2$$

for all n.

23. Let n be an integer with $n \geq 2$. Prove that

$$\prod_{k=1}^{n} \tan\left[\frac{\pi}{3}\left(1 + \frac{3^k}{3^n - 1}\right)\right] = \prod_{k=1}^{n} \cot\left[\frac{\pi}{3}\left(1 - \frac{3^k}{3^n - 1}\right)\right].$$

24. Let $P_2(x) = x^2 - 2$. Find all sequences of polynomials $\{P_k(x)\}_{k=1}^{\infty}$ such that $P_k(x)$ is monic (that is, with leading coefficient 1), has degree k, and $P_i(P_j(x)) = P_j(P_i(x))$ for all i and j.

25. In triangle ABC, $a \leq b \leq c$. As a function of angle C, determine the conditions under which $a + b - 2R - 2r$ is positive, negative, or zero.

26. Let ABC be a triangle. Points D, E, F are on sides BC, CA, AB, respectively, such that $|DC| + |CE| = |EA| + |AF| = |FB| + |BD|$. Prove that

$$|DE| + |EF| + |FD| \geq \frac{1}{2}(|AB| + |BC| + |CA|).$$

27. Let a and b be positive real numbers. Prove that

$$\frac{1}{\sqrt{1+a^2}} + \frac{1}{\sqrt{1+b^2}} \geq \frac{2}{\sqrt{1+ab}}$$

if either (1) $0 < a, b \leq 1$ or (2) $ab \geq 3$.

28. Let ABC be a nonobtuse triangle such that $|AB| > |AC|$ and $\angle B = 45°$. Let O and I denote the circumcenter and incenter of triangle ABC, respectively. Suppose that $\sqrt{2}|OI| = |AB| - |AC|$. Determine all the possible values of $\sin A$.

29. Let n be a positive integer. Find the real numbers a_0 and $a_{k,\ell}$, $1 \le \ell < k \le n$, such that

$$\frac{\sin^2 nx}{\sin^2 x} = a_0 + \sum_{1\le\ell<k\le n} a_{k,\ell} \cos 2(k-\ell)x$$

for all real numbers x not an integer multiple of π.

30. Let S be the set of all triangles ABC for which

$$5\left(\frac{1}{|AP|} + \frac{1}{|BQ|} + \frac{1}{|CR|}\right) - \frac{3}{\min\{|AP|, |BQ|, |CR|\}} = \frac{6}{r},$$

where r is the inradius and P, Q, and R are the points of tangency of the incircle with sides AB, BC, and CA, respectively. Prove that all triangles in S are isosceles and similar to one another.

31. Let a, b, c be real numbers in the interval $(0, \frac{\pi}{2})$. Prove that

$$\frac{\sin a \sin(a-b)\sin(a-c)}{\sin(b+c)} + \frac{\sin b \sin(b-c)\sin(b-a)}{\sin(c+a)} + \frac{\sin c \sin(c-a)\sin(c-b)}{\sin(a+b)} \ge 0.$$

32. Let ABC be a triangle. Prove that

$$\sin\frac{3A}{2} + \sin\frac{3B}{2} + \sin\frac{3C}{2} \le \cos\frac{A-B}{2} + \cos\frac{B-C}{2} + \cos\frac{C-A}{2}.$$

33. Let $x_1, x_2, \ldots, x_n$, $n \ge 2$, be n distinct real numbers in the interval $[-1, 1]$. Prove that

$$\frac{1}{t_1} + \frac{1}{t_2} + \cdots + \frac{1}{t_n} \ge 2^{n-2},$$

where $t_i = \prod_{j\ne i} |x_j - x_i|$.

34. Let $x_1, \ldots, x_{10}$ be real numbers in the interval $[0, \pi/2]$ such that $\sin^2 x_1 + \sin^2 x_2 + \cdots + \sin^2 x_{10} = 1$. Prove that

$$3(\sin x_1 + \cdots + \sin x_{10}) \le \cos x_1 + \cdots + \cos x_{10}.$$

35. Let $x_1, x_2, \ldots, x_n$ be arbitrary real numbers. Prove the inequality

$$\frac{x_1}{1+x_1^2} + \frac{x_2}{1+x_1^2+x_2^2} + \cdots + \frac{x_n}{1+x_1^2+\cdots+x_n^2} < \sqrt{n}.$$

36. Let $a_0, a_1, \ldots, a_n$ be numbers in the interval $\left(0, \frac{\pi}{2}\right)$ such that

$$\tan\left(a_0 - \frac{\pi}{4}\right) + \tan\left(a_1 - \frac{\pi}{4}\right) + \cdots + \tan\left(a_n - \frac{\pi}{4}\right) \geq n - 1.$$

Prove that

$$\tan a_0 \tan a_1 \cdots \tan a_n \geq n^{n+1}.$$

37. Find all triples of real numbers (a, b, c) such that $a^2 - 2b^2 = 1, 2b^2 - 3c^2 = 1$, and $ab + bc + ca = 1$.

38. Let n be a positive integer, and let θ_i be angles with $0 < \theta_i < 90°$ such that

$$\cos^2\theta_1 + \cos^2\theta_2 + \cdots + \cos^2\theta_n = 1.$$

Prove that

$$\tan\theta_1 + \tan\theta_2 + \cdots + \tan\theta_n \geq (n-1)(\cot\theta_1 + \cot\theta_2 + \cdots + \cot\theta_n).$$

39. One of the two inequalities

$$(\sin x)^{\sin x} < (\cos x)^{\cos x} \quad \text{and} \quad (\sin x)^{\sin x} > (\cos x)^{\cos x}$$

is always true for all real numbers x such that $0 < x < \frac{\pi}{4}$. Identify that inequality and prove your result.

40. Let k be a positive integer. Prove that $\sqrt{k+1} - \sqrt{k}$ is not the real part of the complex number z with $z^n = 1$ for some positive integer n.

41. Let $A_1A_2A_3$ be an acute-angled triangle. Points B_1, B_2, B_3 are on sides A_2A_3, A_3A_1, A_1A_2, respectively. Prove that

$$2(b_1\cos A_1 + b_2\cos A_2 + b_3\cos A_3) \geq a_1\cos A_1 + a_2\cos A_2 + a_3\cos A_3,$$

where $a_i = |A_{i+1}A_{i+2}|$ and $b_i = |B_{i+1}B_{i+2}|$, for $i = 1, 2, 3$ (with indices taken modulo 3; that is, $x_{i+3} = x_i$).

42. Let ABC be a triangle. Let x, y, and z be real numbers, and let n be a positive integer. Prove the following four inequalities.

    (a) $x^2 + y^2 + z^2 \ge 2yz\cos A + 2zx\cos B + 2xy\cos C$.

    (b) $x^2 + y^2 + z^2 \ge 2(-1)^{n+1}(yz\cos nA + zx\cos nB + xy\cos nC)$.

    (c) $yza^2 + zxb^2 + xyc^2 \le R^2(x + y + z)^2$.

    (d) $xa^2 + yb^2 + zc^2 \ge 4[ABC]\sqrt{xy + yz + zx}$.

43. A circle ω is inscribed in a quadrilateral $ABCD$. Let I be the center of ω. Suppose that

$$(|AI| + |DI|)^2 + (|BI| + |CI|)^2 = (|AB| + |CD|)^2.$$

Prove that $ABCD$ is an isosceles trapezoid.

44. Let a, b, and c be nonnegative real numbers such that

$$a^2 + b^2 + c^2 + abc = 4.$$

Prove that

$$0 \le ab + bc + ca - abc \le 2.$$

45. Let s, t, u, v be numbers in the interval $\left(0, \frac{\pi}{2}\right)$ with $s + t + u + v = \pi$. Prove that

$$\frac{\sqrt{2}\sin s - 1}{\cos s} + \frac{\sqrt{2}\sin t - 1}{\cos t} + \frac{\sqrt{2}\sin u - 1}{\cos u} + \frac{\sqrt{2}\sin v - 1}{\cos v} \ge 0.$$

46. Suppose a calculator is broken and the only keys that still work are the sin, cos, tan, $\sin^{-1}$, $\cos^{-1}$, and $\tan^{-1}$ buttons. The display initially shows 0. Given any positive rational number q, show that we can get q to appear on the display panel of the calculator by pressing some finite sequence of buttons. Assume that the calculator does real-number calculations with infinite precision, and that all functions are in terms of radians.

47. Let n be a fixed positive integer. Determine the smallest positive real number λ such that for any $\theta_1, \theta_2, \ldots, \theta_n$ in the interval $\left(0, \frac{\pi}{2}\right)$, if

$$\tan\theta_1 \tan\theta_2 \cdots \tan\theta_n = 2^{n/2},$$

then

$$\cos\theta_1 + \cos\theta_2 + \cdots + \cos\theta_n \le \lambda.$$

48. Let ABC be an acute triangle. Prove that

$$(\sin 2B + \sin 2C)^2 \sin A + (\sin 2C + \sin 2A)^2 \sin B \\ + (\sin 2A + \sin 2B)^2 \sin C \le 12 \sin A \sin B \sin C.$$

49. On the sides of a nonobtuse triangle ABC are constructed externally a square P_4, a regular m-sided polygon P_m, and a regular n-sided polygon P_n. The centers of the square and the two polygons form an equilateral triangle. Prove that $m = n = 6$, and find the angles of triangle ABC.

50. Let ABC be an acute triangle. Prove that

$$\left(\frac{\cos A}{\cos B}\right)^2 + \left(\frac{\cos B}{\cos C}\right)^2 + \left(\frac{\cos C}{\cos A}\right)^2 + 8\cos A\cos B\cos C \ge 4.$$

51. For any real number x and any positive integer n, prove that

$$\left|\sum_{k=1}^{n} \frac{\sin kx}{k}\right| \le 2\sqrt{\pi}.$$

4

Solutions to Introductory Problems

1. [AMC12 1999] Let x be a real number such that $\sec x - \tan x = 2$. Evaluate $\sec x + \tan x$.

Solution: Note that

$$1 = \sec^2 x - \tan^2 x = (\sec x + \tan x)(\sec x - \tan x).$$

Hence $\sec x + \tan x = \frac{1}{2}$.

2. Let $0° < \theta < 45°$. Arrange

$$t_1 = (\tan\theta)^{\tan\theta}, \qquad t_2 = (\tan\theta)^{\cot\theta},$$
$$t_3 = (\cot\theta)^{\tan\theta}, \qquad t_4 = (\cot\theta)^{\cot\theta},$$

in decreasing order.

Solution: For $a > 1$, the function $y = a^x$ is an increasing function. For $0° < \theta < 45°$, $\cot\theta > 1 > \tan\theta > 0$. Thus $t_3 < t_4$.

For $a < 1$, the function $y = a^x$ is a decreasing function. Thus $t_1 > t_2$.

Again, by $\cot\theta > 1 > \tan\theta > 0$, we have $t_1 < 1 < t_3$. Hence $t_4 > t_3 > t_1 > t_2$.

3. Compute
   (a) $\sin\frac{\pi}{12}$, $\cos\frac{\pi}{12}$, and $\tan\frac{\pi}{12}$;
   (b) $\cos^4\frac{\pi}{24} - \sin^4\frac{\pi}{24}$;
   (c) $\cos 36^\circ - \cos 72^\circ$; and
   (d) $\sin 10^\circ \sin 50^\circ \sin 70^\circ$.

Solution:

(a) By the **double-angle** and **addition and subtraction formulas**, we have

$$\cos\frac{\pi}{12} = \cos\left(\frac{\pi}{3} - \frac{\pi}{4}\right) = \cos\frac{\pi}{3}\cos\frac{\pi}{4} + \sin\frac{\pi}{3}\sin\frac{\pi}{4}$$
$$= \frac{1}{2}\cdot\frac{\sqrt{2}}{2} + \frac{\sqrt{3}}{2}\cdot\frac{\sqrt{2}}{2} = \frac{\sqrt{2}+\sqrt{6}}{4}.$$

Similarly, we can show that $\sin\frac{\pi}{12} = \frac{\sqrt{6}-\sqrt{2}}{4}$. It follows that $\tan\frac{\pi}{12} = \frac{\sqrt{6}-\sqrt{2}}{\sqrt{6}+\sqrt{2}} = 2 - \sqrt{3}$.

(b) By the double-angle and addition and subtraction formulas, we obtain

$$\cos^4\frac{\pi}{24} - \sin^4\frac{\pi}{24} = \left(\cos^2\frac{\pi}{24} + \sin^2\frac{\pi}{24}\right)\left(\cos^2\frac{\pi}{24} - \sin^2\frac{\pi}{24}\right)$$
$$= 1\cdot\cos\frac{\pi}{12} = \frac{\sqrt{2}+\sqrt{6}}{4}.$$

(c) Note that

$$\cos 36^\circ - \cos 72^\circ = \frac{2(\cos 36^\circ - \cos 72^\circ)(\cos 36^\circ + \cos 72^\circ)}{2(\cos 36^\circ + \cos 72^\circ)}$$
$$= \frac{2\cos^2 36^\circ - 2\cos^2 72^\circ}{2(\cos 36^\circ + \cos 72^\circ)}.$$

By the double-angle formulas, the above equality becomes

$$\cos 36^\circ - \cos 72^\circ = \frac{\cos 72^\circ + 1 - \cos 144^\circ - 1}{2(\cos 36^\circ + \cos 72^\circ)}$$
$$= \frac{\cos 72^\circ + \cos 36^\circ}{2(\cos 36^\circ + \cos 72^\circ)} = \frac{1}{2}.$$

This fact can also be seen in an isosceles triangleABC with $AB = AC$, $BC = 1$, and $A = 36^\circ$. Point D lies on side AC with $\angle ABD = \angle DBC$. We leave it to the reader to show that $BC = BD = AD = 1$, $AB = 2\cos 36$, and $CD = 2\cos 72$, from which the desired result follows.

(d) Applying the double-angle formulas again gives

$$\begin{aligned} 8\sin 20^\circ \sin 10^\circ \sin 50^\circ \sin 70^\circ &= 8\sin 20^\circ \cos 20^\circ \cos 40^\circ \cos 80^\circ \\ &= 4\sin 40^\circ \cos 40^\circ \cos 80^\circ \\ &= 2\sin 80^\circ \cos 80^\circ \\ &= \sin 160^\circ = \sin 20^\circ. \end{aligned}$$

Consequently,

$$\sin 10^\circ \sin 50^\circ \sin 70^\circ = \frac{1}{8}.$$

4. [AMC12P 2002] Simplify the expression

$$\sqrt{\sin^4 x + 4\cos^2 x} - \sqrt{\cos^4 x + 4\sin^2 x}.$$

Solution: The given expression is equal to

$$\begin{aligned} &\sqrt{\sin^4 x + 4(1-\sin^2 x)} - \sqrt{\cos^4 x + 4(1-\cos^2 x)} \\ &\quad = \sqrt{(2-\sin^2 x)^2} - \sqrt{(2-\cos^2 x)^2} = (2-\sin^2 x) - (2-\cos^2 x) \\ &\quad = \cos^2 x - \sin^2 x = \cos 2x. \end{aligned}$$

5. Prove that

$$1 - \cot 23^\circ = \frac{2}{1-\cot 22^\circ}.$$

First Solution: We will show that

$$(1-\cot 23^\circ)(1-\cot 22^\circ) = 2.$$

Indeed, by the **addition and subtraction formulas**, we obtain

$$\begin{aligned} (1-\cot 23^\circ)(1-\cot 22^\circ) &= \left(1 - \frac{\cos 23^\circ}{\sin 23^\circ}\right)\left(1 - \frac{\cos 22^\circ}{\sin 22^\circ}\right) \\ &= \frac{\sin 23^\circ - \cos 23^\circ}{\sin 23^\circ} \cdot \frac{\sin 22^\circ - \cos 22^\circ}{\sin 22^\circ} \\ &= \frac{\sqrt{2}\sin(23^\circ - 45^\circ)\sqrt{2}\sin(22^\circ - 45^\circ)}{\sin 23^\circ \cdot \sin 22^\circ} \\ &= \frac{2\sin(-22^\circ)\sin(-23^\circ)}{\sin 23^\circ \sin 22^\circ} \\ &= \frac{2\sin 22^\circ \sin 23^\circ}{\sin 23^\circ \sin 22^\circ} = 2. \end{aligned}$$

Second Solution: Note that by the addition and subtraction formula, we have

$$\frac{\cot 22^\circ \cot 23^\circ - 1}{\cot 22^\circ + \cot 23^\circ} = \cot(22^\circ + 23^\circ) = \cot 45^\circ = 1.$$

Hence $\cot 22^\circ \cot 23^\circ - 1 = \cot 22^\circ + \cot 23^\circ$, and so

$$1 - \cot 22^\circ - \cot 23^\circ + \cot 22^\circ \cot 23^\circ = 2,$$

that is, $(1 - \cot 23^\circ)(1 - \cot 22^\circ) = 2$, as desired.

6. Find all x in the interval $\left(0, \frac{\pi}{2}\right)$ such that

$$\frac{\sqrt{3} - 1}{\sin x} + \frac{\sqrt{3} + 1}{\cos x} = 4\sqrt{2}.$$

Solution: From Problem 3(a), we have $\cos \frac{\pi}{12} = \frac{\sqrt{2}+\sqrt{6}}{4}$ and $\sin \frac{\pi}{12} = \frac{\sqrt{6}-\sqrt{2}}{4}$. Write the given equation as

$$\frac{\frac{\sqrt{3}-1}{4}}{\sin x} + \frac{\frac{\sqrt{3}+1}{4}}{\cos x} = \sqrt{2},$$

or

$$\frac{\sin \frac{\pi}{12}}{\sin x} + \frac{\cos \frac{\pi}{12}}{\cos x} = 2.$$

Clearing the denominator gives

$$\sin \frac{\pi}{12} \cos x + \cos \frac{\pi}{12} \sin x = 2 \sin x \cos x,$$

or $\sin\left(\frac{\pi}{12} + x\right) = \sin 2x$. We obtain $\frac{\pi}{12} + x = 2x$ and $\frac{\pi}{12} + x = \pi - 2x$, implying that $x = \frac{\pi}{12}$ and $x = \frac{11\pi}{36}$. Both solutions satisfy the given condition.

7. Region $\mathcal{R}$ contains all the points (x, y) such that $x^2 + y^2 \le 100$ and $\sin(x + y) \ge 0$. Find the area of region $\mathcal{R}$.

Solution: Let $\mathcal{C}$ denote the disk (x, y) with $x^2 + y^2 \le 100$. Because $\sin(x + y) = 0$ if and only if $x + y = k\pi$ for integers k, disk $\mathcal{C}$ has been cut by parallel lines $x+y = k\pi$, and in between those lines there are regions containing points (x, y) with either $\sin(x + y) > 0$ or $\sin(x + y) < 0$. Since $\sin(-x - y) = -\sin(x + y)$, the regions containing points (x, y) with $\sin(x + y) > 0$ are symmetric with respect to the origin to the regions containing points (x, y) with $\sin(x + y) < 0$. Thus, as indicated in Figure 4.1, the area of region $\mathcal{R}$ is half the area of disk $\mathcal{C}$, that is, 50π.

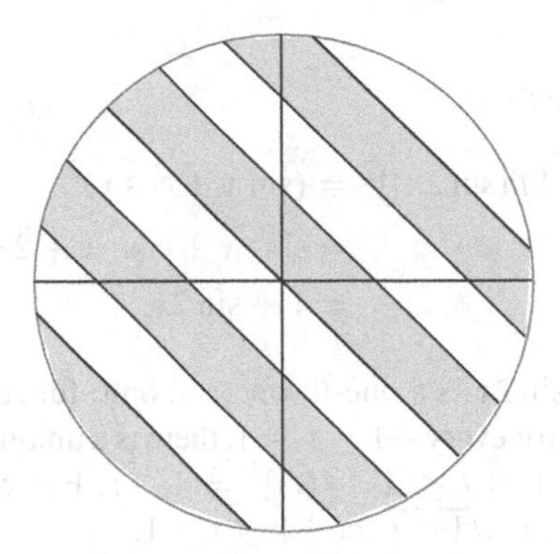

Figure 4.1.

8. In triangle ABC, show that

$$\sin\frac{A}{2} \le \frac{a}{b+c}.$$

Solution: By the **extended law of sines**, we have

$$\frac{a}{b+c} = \frac{\sin A}{\sin B + \sin C}.$$

Applying the **double-angle formulas** and **sum-to-product formulas** in the above relation gives

$$\frac{a}{b+c} = \frac{2\sin\frac{A}{2}\cos\frac{A}{2}}{2\sin\frac{B+C}{2}\cos\frac{B-C}{2}} = \frac{\sin\frac{A}{2}}{\cos\frac{B-C}{2}} \ge \sin\frac{A}{2},$$

by noting that $0 < \cos\frac{B-C}{2} \le 1$, because $0 \le |B - C| < 180^\circ$.

Note: By symmetry, we have analogous formulas

$$\sin\frac{B}{2} \le \frac{b}{c+a} \quad \text{and} \quad \sin\frac{C}{2} \le \frac{c}{a+b}.$$

9. Let I denote the interval $[-\frac{\pi}{4}, \frac{\pi}{4}]$. Determine the function f defined on the interval $[-1, 1]$ such that $f(\sin 2x) = \sin x + \cos x$ and simplify $f(\tan^2 x)$ for x in the interval I.

Solution: Note that

$$\begin{aligned}
[f(\sin 2x)]^2 &= (\sin x + \cos x)^2 \\
&= \sin^2 x + \cos^2 x + 2 \sin x \cos x \\
&= 1 + \sin 2x.
\end{aligned}$$

Note also that $\sin 2x$ is a one-to-one and onto function from I to the interval $[-1, 1]$, that is, for every $-1 \le t \le 1$, there is a unique x in I such that $\sin 2x = t$. Hence, for $-1 \le t \le 1$, $[f(t)]^2 = 1 + t$. For x in I, $\sin x + \cos x \ge 0$. Therefore, $f(t) = \sqrt{1+t}$ for $-1 \le t \le 1$.

For $\frac{\pi}{4} \le x \le \frac{\pi}{4}$, $-1 \le \tan x \le 1$, and so $0 \le \tan^2 x \le 1$. Thus,

$$f(\tan^2 x) = \sqrt{1 + \tan^2 x} = \sec x.$$

10. Let

$$f_k(x) = \frac{1}{k}(\sin^k x + \cos^k x)$$

for $k = 1, 2, \ldots$. Prove that

$$f_4(x) - f_6(x) = \frac{1}{12}$$

for all real numbers x.

Solution: We need to show that

$$3(\sin^4 x + \cos^4 x) - 2(\sin^6 x + \cos^6 x) = 1$$

for all real numbers x. Indeed, the left-hand side is equal to

$$\begin{aligned}
&3[(\sin^2 x + \cos^2 x)^2 - 2 \sin^2 x \cos^2 x] \\
&\quad - 2(\sin^2 x + \cos^2 x)(\sin^4 x - \sin^2 x \cos^2 x + \cos^4 x) \\
&\qquad = 3 - 6 \sin^2 x \cos^2 x - 2[(\sin^2 x + \cos^2 x)^2 - 3 \sin^2 x \cos^2 x] \\
&\qquad = 3 - 2 = 1.
\end{aligned}$$

11. [AIME 2004, by Jonathan Kane] A circle of radius 1 is randomly placed in a 15×36 rectangle $ABCD$ so that the circle lies completely within the rectangle. Compute the probability that the circle will not touch diagonal AC.

Note: In order for the circle to lie completely within the rectangle, the center of the circle must lie in a rectangle that is $(15-2) \times (36-2)$, or 13×34. The requested probability is equal to the probability that the distance from the circle's center to the diagonal AC is greater than 1, which equals the probability that the distance from a randomly selected point in the 13×34 rectangle to each side of triangles ABC and CDA is greater than 1. Let $|AB| = 36$ and $|BC| = 15$ (and so $|AC| = 39$). Draw three segments that are 1 unit away from each side of triangle ABC and whose endpoints are on the sides. Let E, F, and G be the three points of intersection nearest to A, B, and C, respectively, of the three segments. Because the corresponding sides of triangle ABC and EFG are parallel, the two triangles are similar to each other. The desired probability is equal to

$$\frac{2[EFG]}{13 \cdot 34} = \left(\frac{|EF|}{|AB|}\right)^2 \cdot \frac{2[ABC]}{13 \cdot 34} = \left(\frac{|EF|}{|AB|}\right)^2 \cdot \frac{15 \cdot 36}{13 \cdot 34} = \left(\frac{|EF|}{|AB|}\right)^2 \cdot \frac{270}{221}.$$

Because E is equidistant from sides AB and AC, E lies on the bisector of $\angle CAB$. Similarly, F and G lie on the bisectors of $\angle ABC$ and $\angle BCA$, respectively. Hence lines AE, BF, and CG meet I, the incenter of triangle ABC.

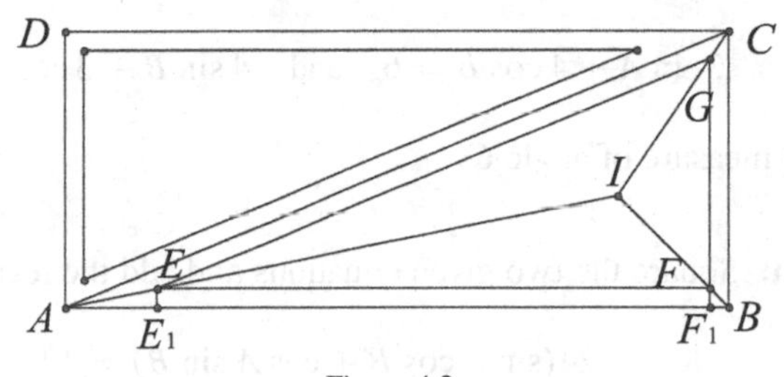

Figure 4.2.

First Solution: Let E_1 and F_1 be the feet of the perpendiculars from E and F to segment AB, respectively. Then $|EF| = |E_1F_1|$. It is not difficult to see that $|BF_1| = |FF_1| = |EE_1| = 1$. Set $\theta = \angle EAB$. Then $\angle CAB = 2\theta$, $\sin 2\theta = \frac{5}{12}$, $\cos 2\theta = \frac{12}{13}$, and $\tan 2\theta = \frac{5}{12}$. By either the **double-angle formulas** or the **half-angle formulas**,

$$\tan 2\theta = \frac{2\tan\theta}{1-\tan^2\theta} \quad \text{or} \quad \tan\theta = \frac{1-\cos 2\theta}{\sin 2\theta},$$

and we obtain $\tan\theta = \frac{1}{5}$. It follows that $\frac{|EE_1|}{|AE_1|} = \tan\theta = \frac{1}{5}$, or $|AE_1| = 5$. Consequently, $|EF| = |E_1F_1| = 30$. Hence $\frac{m}{n} = \left(\frac{30}{36}\right)^2 \cdot \frac{270}{221} = \frac{375}{442}$, and $m + n = 817$.

Second Solution: Set $A = (0, 0)$, $B = (36, 0)$, and $C = (36, 15)$. Because E lies on the angle bisector of $\angle CAB$, $\overrightarrow{AE}$ has the same slope as $|\overrightarrow{AB}|\overrightarrow{AC} + |\overrightarrow{AC}|\overrightarrow{AB} = 36[36, 15] + 39[36, 0] = [75 \cdot 36, 36 \cdot 15] = 36 \cdot 15[5, 1]$; that is, the slope of line AE is $\frac{1}{5}$. Consequently, $|EE_1| = 5$, and the rest of the solution proceeds like that of the first solution.

Third Solution: Because the corresponding sides of triangles ABC and EFG are parallel, it follows that I is also the incenter of triangle EFG and that the triangles are **homothetic** (with I as the center). If r is the inradius of triangle ABC, then $r - 1$ is the inradius of triangle EFG; that is, the ratio of the similarity between triangles EFG and ABC is $\frac{r-1}{r}$. Hence the desired probability is $\left(\frac{r-1}{r}\right)^2 \cdot \frac{270}{221}$.

Note that $r(|AB|+|BC|+|CA|) = 2([AIB]+[BIC]+[CIA]) = 2[ABC] = |AB| \cdot |BC|$. Solving the last equation gives $r = 6$, and so $\left(\frac{5}{6}\right)^2 \cdot \frac{270}{221} = \frac{375}{442}$.

12. [AMC12 1999] In triangle ABC,

$$3\sin A + 4\cos B = 6 \quad \text{and} \quad 4\sin B + 3\cos A = 1.$$

Find the measure of angle C.

Solution: Square the two given equations and add the results to obtain

$$24(\sin A\cos B + \cos A\sin B) = 12,$$

or $\sin(A+B) = \frac{1}{2}$. Because $C = 180^\circ - A - B$, we have $\sin C = \sin(A+B) = \frac{1}{2}$, implying that either $C = 30^\circ$ or $C = 150^\circ$. But if $C = 150^\circ$, $A < 30^\circ$, so $3\sin A + 4\cos B < \frac{3}{2} + 4 < 6$, a contradiction. Hence the answer is $C = 30^\circ$.

13. Prove that

$$\tan 3a - \tan 2a - \tan a = \tan 3a \tan 2a \tan a$$

for all $a \neq \frac{k\pi}{2}$, where k is in $\mathbb{Z}$.

Solution: The equality is equivalent to

$$\tan 3a(1 - \tan 2a \tan a) = \tan 2a + \tan a,$$

or

$$\tan 3a = \frac{\tan 2a + \tan a}{1 - \tan 2a \tan a}.$$

That is, $\tan 3a = \tan(2a + a)$, which is evident.

Note: More generally, if a_1, a_2, a_3 are real numbers different from $\frac{k\pi}{2}$, where k is in $\mathbb{Z}$, such that $a_1 + a_2 + a_3 = 0$, then the relation

$$\tan a_1 + \tan a_2 + \tan a_3 = \tan a_1 \tan a_2 \tan a_3$$

holds. The proof of this relation is similar to the proofs of Problems 13 and 20. We leave the proof as an exercise for the reader.

14. Let a, b, c, d be numbers in the interval $[0, \pi]$ such that

$$\sin a + 7 \sin b = 4(\sin c + 2 \sin d),$$
$$\cos a + 7 \cos b = 4(\cos c + 2 \cos d).$$

Prove that $2 \cos(a - d) = 7 \cos(b - c)$.

Solution: Rewrite the two given equalities as

$$\sin a - 8 \sin d = 4 \sin c - 7 \sin b,$$
$$\cos a - 8 \cos d = 4 \cos c - 7 \cos b.$$

By squaring the last two equalities and adding them, we obtain

$$1+64-16(\cos a \cos d+\sin a \sin d) = 16+49-56(\cos b \cos c+\sin b \sin c),$$

and the conclusion follows from the **addition formulas**.

15. Express

$$\sin(x - y) + \sin(y - z) + \sin(z - x)$$

as a monomial.

Solution: By the **sum-to-product formulas**, we have

$$\sin(x-y)+\sin(y-z)=2\sin\frac{x-z}{2}\cos\frac{x+z-2y}{2}.$$

By the **double-angle formulas**, we have

$$\sin(z-x)=2\sin\frac{z-x}{2}\cos\frac{z-x}{2}.$$

Thus,

$$\begin{aligned}&\sin(x-y)+\sin(y-z)+\sin(z-x)\\&=2\sin\frac{x-z}{2}\left[\cos\frac{x+z-2y}{2}-\cos\frac{z-x}{2}\right]\\&=-4\sin\frac{x-z}{2}\sin\frac{z-y}{2}\sin\frac{x-y}{2}\\&=-4\sin\frac{x-y}{2}\sin\frac{y-z}{2}\sin\frac{z-x}{2}\end{aligned}$$

by the sum-to-product formulas.

Note: In exactly the same way, we can show that if a, b, and c are real numbers with $a+b+c=0$, then

$$\sin a+\sin b+\sin c=-4\sin\frac{a}{2}\sin\frac{b}{2}\sin\frac{c}{2}.$$

In Problem 15, we have $a=x-y$, $b=y-z$, and $c=z-x$.

16. Prove that

$$(4\cos^2 9^\circ-3)(4\cos^2 27^\circ-3)=\tan 9^\circ.$$

Solution: We have $\cos 3x=4\cos^3 x-3\cos x$, so $4\cos^2 x-3=\frac{\cos 3x}{\cos x}$ for all $x\neq(2k+1)\cdot 90^\circ$, $k\in\mathbb{Z}$. Thus

$$\begin{aligned}(4\cos^2 9^\circ-3)(4\cos^2 27^\circ-3)&=\frac{\cos 27^\circ}{\cos 9^\circ}\cdot\frac{\cos 81^\circ}{\cos 27^\circ}=\frac{\cos 81^\circ}{\cos 9^\circ}\\&=\frac{\sin 9^\circ}{\cos 9^\circ}=\tan 9^\circ,\end{aligned}$$

as desired.

17. Prove that

$$\left(1+\frac{a}{\sin x}\right)\left(1+\frac{b}{\cos x}\right) \geq \left(1+\sqrt{2ab}\right)^2$$

for all real numbers a, b, x with $a, b \geq 0$ and $0 < x < \frac{\pi}{2}$.

Solution: Expanding both sides, the desired inequality becomes

$$1+\frac{a}{\sin x}+\frac{b}{\cos x}+\frac{ab}{\sin x \cos x} > 1+2ab+2\sqrt{2ab}.$$

By the **arithmetic–geometric means inequality**, we obtain

$$\frac{a}{\sin x}+\frac{b}{\cos x} \geq \frac{2\sqrt{ab}}{\sqrt{\sin x \cos x}}.$$

By the **double-angle formulas**, we have $\sin x \cos x = \frac{1}{2}\sin 2x \leq \frac{1}{2}$, and so

$$\frac{2\sqrt{ab}}{\sqrt{\sin x \cos x}} \geq 2\sqrt{2ab}$$

and

$$\frac{ab}{\sin x \cos x} \geq 2ab.$$

Combining the last three inequalities gives the the desired result.

18. In triangle ABC, $\sin A + \sin B + \sin C \leq 1$. Prove that

$$\min\{A+B, B+C, C+A\} < 30°.$$

Solution: Without loss of generality, we assume that $A \geq B \geq C$. We need to prove that $B+C < 30°$. The **law of sines** and the **triangle inequality** ($b+c > a$) imply that $\sin B + \sin C > \sin A$, so $\sin A + \sin B + \sin C > 2\sin A$. It follows that $\sin A < \frac{1}{2}$, and the inequality $A \geq \frac{A+B+C}{3} = 60°$ gives that $A > 150°$; that is, $B+C < 30°$, as desired.

19. Let ABC be a triangle. Prove that

(a)

$$\tan\frac{A}{2}\tan\frac{B}{2}+\tan\frac{B}{2}\tan\frac{C}{2}+\tan\frac{C}{2}\tan\frac{A}{2} = 1;$$

(b)

$$\tan\frac{A}{2}\tan\frac{B}{2}\tan\frac{C}{2}\le\frac{\sqrt{3}}{9}.$$

Solution: By the **addition and subtraction formulas**, we have

$$\tan\frac{A}{2}+\tan\frac{B}{2}=\tan\frac{A+B}{2}\left(1-\tan\frac{A}{2}\tan\frac{B}{2}\right).$$

Because $A+B+C=180^\circ$, $\frac{A+B}{2}=90^\circ-\frac{C}{2}$, and so $\tan\frac{A+B}{2}=\cot\frac{C}{2}$. Thus,

$$\begin{aligned}
&\tan\frac{A}{2}\tan\frac{B}{2}+\tan\frac{B}{2}\tan\frac{C}{2}+\tan\frac{C}{2}\tan\frac{A}{2}\\
&=\tan\frac{A}{2}\tan\frac{B}{2}+\tan\frac{C}{2}\cot\frac{C}{2}\left(1-\tan\frac{A}{2}\tan\frac{B}{2}\right)\\
&=\tan\frac{A}{2}\tan\frac{B}{2}+1-\tan\frac{A}{2}\tan\frac{B}{2}=1,
\end{aligned}$$

establishing (a).

By the **arithmetic–geometric means inequality**, we have

$$\begin{aligned}
1&=\tan\frac{A}{2}\tan\frac{B}{2}+\tan\frac{B}{2}\tan\frac{C}{2}+\tan\frac{C}{2}\tan\frac{A}{2}\\
&\ge 3\sqrt[3]{\left(\tan\frac{A}{2}\tan\frac{B}{2}\tan\frac{C}{2}\right)^2},
\end{aligned}$$

from which (b) follows.

Note: An equivalent form of (a) is

$$\cot\frac{A}{2}+\cot\frac{B}{2}+\cot\frac{C}{2}=\cot\frac{A}{2}\cot\frac{B}{2}\cot\frac{C}{2}.$$

20. Let ABC be an acute-angled triangle. Prove that

(a) $\tan A+\tan B+\tan C=\tan A\tan B\tan C$;

(b) $\tan A\tan B\tan C\ge 3\sqrt{3}$.

Solution: Note that because of the condition $A, B, C \neq 90°$, all the above expressions are well defined.

The proof of the identity in part (a) is similar to that of Problem 19(a). By the **arithmetic–geometric means inequality**,

$$\tan A + \tan B + \tan C \geq 3\sqrt[3]{\tan A \tan B \tan C}.$$

By (a), we have

$$\tan A \tan B \tan C \geq 3\sqrt[3]{\tan A \tan B \tan C},$$

from which (b) follows.

Note: Indeed, the identity in (a) holds for all angles A, B, C with $A+B+C = m\pi$ and $A, B, C \neq \frac{k\pi}{2}$, where k and m are in $\mathbb{Z}$.

21. Let ABC be a triangle. Prove that

$$\cot A \cot B + \cot B \cot C + \cot C \cot A = 1.$$

Conversely, prove that if x, y, z are real numbers with $xy + yz + zx = 1$, then there exists a triangle ABC such that $\cot A = x$, $\cot B = y$, and $\cot C = z$.

Solution: If ABC is a right triangle, then without loss of generality, assume that $A = 90°$. Then $\cot A = 0$ and $B + C = 90°$, and so $\cot B \cot C = 1$, implying the desired result.

If $A, B, C \neq 90°$, then $\tan A \tan B \tan C$ is well defined. Multiplying both sides of the desired identity by $\tan A \tan B \tan C$ reduces the desired result to Introductory Problem 20(a).

The second claim is true because $\cot x$ is a bijective function from the interval $(0°, 180°)$ to $(-\infty, \infty)$.

22. Let ABC be a triangle. Prove that

$$\sin^2 \frac{A}{2} + \sin^2 \frac{B}{2} + \sin^2 \frac{C}{2} + 2 \sin \frac{A}{2} \sin \frac{B}{2} \sin \frac{C}{2} = 1.$$

Conversely, prove that if x, y, z are positive real numbers such that

$$x^2 + y^2 + z^2 + 2xyz = 1,$$

then there is a triangle ABC such that $x = \sin\frac{A}{2}$, $y = \sin\frac{B}{2}$, and $z = \sin\frac{C}{2}$.

Solution: Solving the second given equation as a quadratic in x gives

$$x = \frac{-2yz + \sqrt{4y^2z^2 - 4(y^2 + z^2 - 1)}}{2} = -yz + \sqrt{(1 - y^2)(1 - z^2)}.$$

We make the trigonometric substitution $y = \sin u$ and $z = \sin v$, where $0^\circ < u, v < 90^\circ$. Then

$$x = -\sin u \sin v + \cos u \cos v = \cos(u + v).$$

Set $u = \frac{B}{2}$, $v = \frac{C}{2}$, and $A = 180^\circ - B - C$. Because $1 \geq y^2 + z^2 = \sin^2\frac{B}{2} + \sin^2\frac{C}{2}$, $\cos^2\frac{B}{2} \geq \sin^2\frac{C}{2}$. Because $0^\circ < \frac{B}{2}, \frac{C}{2} < 90^\circ$, $\cos\frac{B}{2} > \sin\frac{C}{2} = \cos\left(90^\circ - \frac{C}{2}\right)$, implying that $\frac{B}{2} < 90^\circ - \frac{C}{2}$, or $B + C < 180^\circ$. Then $x = \cos(u + v) = \sin\frac{A}{2}$, $y = \sin\frac{B}{2}$, and $z = \sin\frac{C}{2}$, where A, B, and C are the angles of a triangle.

If ABC is a triangle, all the above steps can be reversed to obtain the first given identity.

23. Let ABC be a triangle. Prove that

(a) $\sin\dfrac{A}{2}\sin\dfrac{B}{2}\sin\dfrac{C}{2} \leq \dfrac{1}{8}$;

(b) $\sin^2\dfrac{A}{2} + \sin^2\dfrac{B}{2} + \sin^2\dfrac{C}{2} \geq \dfrac{3}{4}$;

(c) $\cos^2\dfrac{A}{2} + \cos^2\dfrac{B}{2} + \cos^2\dfrac{C}{2} \leq \dfrac{9}{4}$;

(d) $\cos\dfrac{A}{2}\cos\dfrac{B}{2}\cos\dfrac{C}{2} \leq \dfrac{3\sqrt{3}}{8}$;

(e) $\csc\dfrac{A}{2} + \csc\dfrac{A}{2} + \csc\dfrac{A}{2} \geq 6$.

Solution: By Problem 8, we have

$$\sin\frac{A}{2}\sin\frac{B}{2}\sin\frac{C}{2} \leq \frac{abc}{(a + b)(b + c)(c + a)}.$$

The **arithmetic–geometric means inequality** yields

$$(a + b)(b + c)(c + a) \geq (2\sqrt{ab})(2\sqrt{bc})(2\sqrt{ca}) = 8abc.$$

Combining the last two equalities gives part (a).

Part (b) then follows from (a) and Problem 22. Part (c) then follows from part (b) by noting that $1 - \sin^2 x = \cos^2 x$. Finally, by (c) and by the arithmetic–geometric means inequality, we have

$$\frac{9}{4} \geq \cos^2 \frac{A}{2} + \cos^2 \frac{B}{2} + \cos^2 \frac{C}{2} \geq 3\sqrt[3]{\cos^2 \frac{A}{2} \cos^2 \frac{B}{2} \cos^2 \frac{C}{2}},$$

implying (d).

Again by Problem 8, we have

$$\csc \frac{A}{2} \geq \frac{b+c}{a} = \frac{b}{a} + \frac{c}{a}$$

and analogous formulas for $\csc \frac{B}{2}$ and $\csc \frac{C}{2}$. Then part (e) follows routinely from the arithmetic–geometric means inequality.

Note: We present another approach to part (a). Note that $\sin \frac{A}{2}$, $\sin \frac{B}{2}$, $\sin \frac{C}{2}$ are all positive. Let $t = \sqrt[3]{\sin \frac{A}{2} \sin \frac{B}{2} \sin \frac{C}{2}}$. It suffices to show that $t \leq \frac{1}{2}$. By the arithmetic–geometric means inequality, we have

$$\sin^2 \frac{A}{2} + \sin^2 \frac{B}{2} + \sin^2 \frac{C}{2} \geq 3t^2.$$

By Problem 22, we have $3t^2 + 2t^3 \leq 1$. Thus,

$$0 \geq 2t^3 + 3t^2 - 1 = (t+1)(2t^2 + t - 1) = (t+1)^2(2t - 1).$$

Consequently, $t \leq \frac{1}{2}$, establishing (a).

24. In triangle ABC, show that

    (a) $\sin 2A + \sin 2B + \sin 2C = 4 \sin A \sin B \sin C$;

    (b) $\cos 2A + \cos 2B + \cos 2C = -1 - 4 \cos A \cos B \cos C$;

    (c) $\sin^2 A + \sin^2 B + \sin^2 C = 2 + 2 \cos A \cos B \cos C$;

    (d) $\cos^2 A + \cos^2 B + \cos^2 C + 2 \cos A \cos B \cos C = 1$.

    Conversely, if x, y, z are positive real numbers such that

    $$x^2 + y^2 + z^2 + 2xyz = 1,$$

    show that there is an acute triangle ABC such that $x = \cos A$, $y = \cos B$, $C = \cos C$.

Solution: Parts (c) and (d) follow immediately from (b) because $\cos 2x = 1 - 2\sin^2 x = 2\cos^2 x - 1$. Thus we show only (a) and (b).

(a) Applying the **sum-to-product formulas** and the fact that $A + B + C = 180^\circ$, we find that

$$\begin{aligned}\sin 2A + \sin 2B + \sin 2C &= 2\sin(A+B)\cos(A-B) + \sin 2C \\ &= 2\sin C\cos(A-B) + 2\sin C\cos C \\ &= 2\sin C[\cos(A-B) - \cos(A+B)] \\ &= 2\sin C \cdot [-2\sin A\sin(-B)] \\ &= 4\sin A\sin B\sin C,\end{aligned}$$

establishing (a).

(b) By the sum-to-product formulas, we have $\cos 2A + \cos 2B = 2\cos(A+B)\cos(A-B) = -2\cos C\cos(A-B)$, because $A + B + C = 180^\circ$. Note that $\cos 2C + 1 = 2\cos^2 C$. It suffices to show that

$$-2\cos C(\cos(A-B) - \cos C) = -4\cos A\cos B\cos C,$$

or $\cos C(\cos(A-B) + \cos(A+B)) = 2\cos A\cos B\cos C$, which is evident by the sum-to-product formula $\cos(A-B) + \cos(A+B) = 2\cos A\cos B$.

From the given equality, we have $1 \ge x^2$, $1 \ge y^2$, and thus we may set $x = \cos A$, $y = \cos B$, where $0^\circ \le A, B \le 90^\circ$. Because $x^2 + y^2 + z^2 + 2xyz$ is an increasing function of z, there is at most one nonnegative value c such that the given equality holds. We know that one solution to this equality is $z = \cos C$, where $C = 180^\circ - A - B$. Because $\cos^2 A + \cos^2 B = x^2 + y^2 \le 1$, we know that $\cos^2 B \le \sin^2 A$. Because $0^\circ < A, B \le 90^\circ$, we have $\cos B \le \sin A = \cos(90^\circ - A)$, implying that $A + B \ge 90^\circ$. Thus, $C \le 90^\circ$ and $\cos C \ge 0$. Therefore, we must have $z = \cos C$, as desired.

Note: Nevertheless, we present a cool proof of part (d). Consider the system of equations

$$\begin{aligned}-x + (\cos B)y + (\cos C)z &= 0 \\ (\cos B)x - y + (\cos A)z &= 0 \\ (\cos C)x + (\cos A)y - z &= 0.\end{aligned}$$

Using the **addition and subtraction formulas**, one can easily see that $(x, y, z) = (\sin A, \sin C, \sin B)$ is a nontrivial solution. Hence the determinant of the system is 0; that is,

$$0 = \begin{vmatrix} -1 & \cos B & \cos C \\ \cos B & -1 & \cos A \\ \cos C & \cos A & -1 \end{vmatrix}$$
$$= -1 + 2\cos A \cos B \cos C + \cos^2 A + \cos^2 B + \cos^2 C,$$

as desired.

25. In triangle ABC, show that

(a) $4R = \dfrac{abc}{[ABC]}$;

(b) $2R^2 \sin A \sin B \sin C = [ABC]$;

(c) $2R \sin A \sin B \sin C = r(\sin A + \sin B + \sin C)$;

(d) $r = 4R \sin \dfrac{A}{2} \sin \dfrac{B}{2} \sin \dfrac{C}{2}$;

(e) $a \cos A + b \cos B + c \cos C = \dfrac{abc}{2R^2}$.

Solution: By the **extended law of sines**,

$$R = \frac{a}{2\sin A} = \frac{abc}{2bc \sin A} = \frac{abc}{4[ABC]},$$

establishing (a). By the same token, we have

$$2R^2 \sin A \sin B \sin C = \frac{1}{2} \cdot (2R \sin A)(2R \sin B)(\sin C)$$
$$= \frac{1}{2} ab \sin C = [ABC],$$

which is (b).

Note that

$$2[ABC] = bc \sin A = (a + b + c)r.$$

By the extended law of sines, we obtain

$$4R^2 \sin A \sin B \sin C = bc \sin A = r(a + b + c)$$
$$= 2rR(\sin A + \sin B + \sin C),$$

from which (c) follows.

By the **law of cosines,**

$$\cos A = \frac{b^2 + c^2 - a^2}{2bc}.$$

Hence, by the **half-angle formulas**, we have

$$\begin{aligned}\sin^2 \frac{A}{2} &= \frac{1 - \cos A}{2} = \frac{1}{2} - \frac{b^2 + c^2 - a^2}{4bc} = \frac{a^2 - (b^2 + c^2 - 2bc)}{4bc} \\ &= \frac{a^2 - (b - c)^2}{4bc} = \frac{(a - b + c)(a + b - c)}{4bc} \\ &= \frac{(2s - 2b)(2s - 2c)}{4bc} = \frac{(s - b)(s - c)}{bc},\end{aligned}$$

where $2s = a + b + c$ is the perimeter of triangle ABC. It follows that

$$\sin \frac{A}{2} = \sqrt{\frac{(s - b)(s - c)}{bc}},$$

and the analogous formulas for $\sin \frac{B}{2}$ and $\sin \frac{C}{2}$. Hence

$$\begin{aligned}\sin \frac{A}{2} \sin \frac{B}{2} \sin \frac{C}{2} &= \frac{(s - a)(s - b)(s - c)}{abc} \\ &= \frac{s(s - a)(s - b)(s - c)}{sabc} = \frac{[ABC]^2}{sabc}\end{aligned}$$

by **Heron's formula**. It follows that

$$\sin \frac{A}{2} \sin \frac{B}{2} \sin \frac{C}{2} = \frac{[ABC]}{s} \cdot \frac{[ABC]}{abc} = r \cdot \frac{1}{4R},$$

from which (d) follows.

Now we prove (e). By the extended law of sines, we have $a \cos A = 2R \sin A \cdot \cos A = R \sin 2A$. Likewise, $b \cos B = R \sin 2B$ and $c \cos C = R \sin 2C$. By (a) and (b), we have

$$4R \sin A \sin B \sin C = \frac{abc}{2R^2}.$$

It suffices to show that

$$\sin 2A + \sin 2B + \sin 2C = 4 \sin A \sin B \sin C,$$

which is Problem 24(a).

26. Let s be the semiperimeter of triangle ABC. Prove that

(a) $s = 4R\cos\frac{A}{2}\cos\frac{B}{2}\cos\frac{C}{2}$;

(b) $s \le \frac{3\sqrt{3}}{2}R$.

Solution: It is well known that $rs = [ABC]$, or $s = \frac{[ABC]}{r}$. By Problem 25 (b) and (d), part (a) follows from

$$s = \frac{R\sin A\sin B\sin C}{2\sin\frac{A}{2}\sin\frac{B}{2}\sin\frac{C}{2}} = 4R\cos\frac{A}{2}\cos\frac{B}{2}\cos\frac{C}{2}$$

by the **double-angle formulas**.

We conclude part (b) from (a) and Problem 23 (d).

27. In triangle ABC, show that

(a) $\cos A + \cos B + \cos C = 1 + 4\sin\frac{A}{2}\sin\frac{B}{2}\sin\frac{C}{2}$;

(b) $\cos A + \cos B + \cos C \le \frac{3}{2}$.

Solution: By the **sum-to-product** and the **double-angle formulas**, we have

$$\cos A + \cos B = 2\cos\frac{A+B}{2}\cos\frac{A-B}{2} = 2\sin\frac{C}{2}\cos\frac{A-B}{2}$$

and

$$1 - \cos C = 2\sin^2\frac{C}{2} = 2\sin\frac{C}{2}\cos\frac{A+B}{2}.$$

It suffices to show that

$$2\sin\frac{C}{2}\left[\cos\frac{A-B}{2} - \cos\frac{A+B}{2}\right] = 4\sin\frac{A}{2}\sin\frac{B}{2}\sin\frac{C}{2},$$

or,

$$\cos\frac{A-B}{2} - \cos\frac{A+B}{2} = 2\sin\frac{A}{2}\sin\frac{B}{2},$$

which follows from the sum-to-product formulas, and hence (a) is established.

Recalling Problem 25 (c), we have

$$\cos A + \cos B + \cos C = 1 + \frac{r}{R}. \qquad (*)$$

Euler's formula states that $|OI|^2 = R^2 - 2Rr$, where O and I are the circumcenter and incenter of triangle ABC. Because $|OI|^2 \geq 0$, we have $R \geq 2r$, or $\frac{r}{R} \leq \frac{1}{2}$, from which (b) follows.

Note: Relation $(*)$ also has a geometric interpretation.

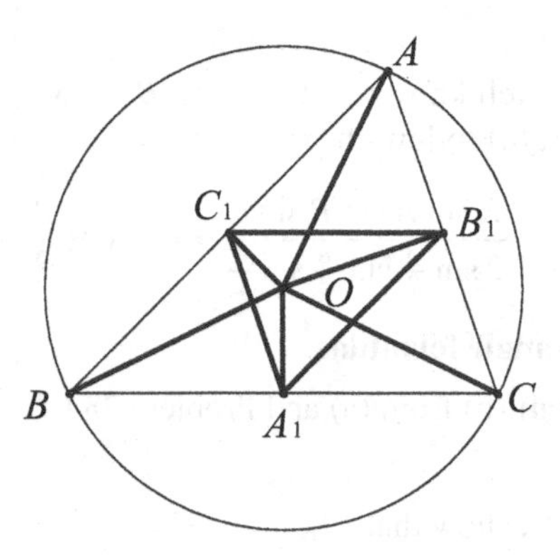

Figure 4.3.

As shown in the Figure 4.3, let O be the circumcenter, and let A_1, B_1, C_1 be the feet of the perpendiculars from O to sides BC, CA, AB, respectively. (Thus A_1, B_1, C_1 are the midpoints of sides BC, CA, AB, respectively.) Because $\angle AOB = 2C$ and triangle AOB is isosceles with $|OA| = |OB| = R$, we have $|OC_1| = R\cos C$. Likewise, $|OB_1| = R\cos B$ and $|OA_1| = R\cos A$. It suffices to show that

$$|OA_1| + |OB_1| + |OC_1| = R + r.$$

Note that $|OA| = |OB| = |OC| = R$ and $|BA_1| = |A_1C|$, $|CB_1| = |B_1A|$, $|AC_1| = |C_1B|$. Hence $|AB| = 2|A_1B_1|$, $|BC| = 2|B_1C_1|$, $|CA| = 2|C_1A_1|$.

Let s denote the semiperimeter of triangle ABC. Applying **Ptolemy's theorem** to cyclic quadrilaterals OA_1CB_1, OB_1AC_1, OC_1BA_1 yields

$$\begin{aligned}
|A_1B_1| \cdot |OC| &= |A_1C| \cdot |OB_1| + |CB_1| \cdot |OA_1|,\\
|B_1C_1| \cdot |OA| &= |B_1A| \cdot |OC_1| + |AC_1| \cdot |OB_1|,\\
|C_1A_1| \cdot |OB| &= |C_1B| \cdot |OA_1| + |BA_1| \cdot |OC_1|.
\end{aligned}$$

Adding the above gives

$$\begin{aligned}
Rs &= |OA_1|(s - |A_1B|) + |OB_1|(s - |B_1C|) + |OC_1|(s - |C_1A|)\\
&= s(|OA_1| + |OB_1| + |OC_1|) - [ABC]\\
&= s(|OA_1| + |OB_1| + |OC_1|) - rs,
\end{aligned}$$

from which our desired result follows.

28. Let ABC be a triangle. Prove that

(a) $\cos A \cos B \cos C \leq \frac{1}{8}$;

(b) $\sin A \sin B \sin C \leq \frac{3\sqrt{3}}{8}$;

(c) $\sin A + \sin B + \sin C \leq \frac{3\sqrt{3}}{2}$.

(d) $\cos^2 A + \cos^2 B + \cos^2 C \geq \frac{3}{4}$;

(e) $\sin^2 A + \sin^2 B + \sin^2 C \leq \frac{9}{4}$;

(f) $\cos 2A + \cos 2B + \cos 2C \geq -\frac{3}{2}$;

(g) $\sin 2A + \sin 2B + \sin 2C \leq \frac{3\sqrt{3}}{2}$.

Solution: For part (a), if triangle ABC is nonacute, the left-hand side of the inequality is nonpositive, and so the inequality is clearly true.

If ABC is acute, then $\cos A$, $\cos B$, $\cos C$ are all positive. To establish (a) and (d), we need only note that the relation between (a) and (d) and Problem 24(d) is similar to that of Problem 23(a) and (b) and Problem 22. (Please see the note after the solution of Problem 23.)

The two inequalities in parts (d) and (e) are equivalent because $\cos^2 x + \sin^2 x = 1$.

By (e) and by the **arithmetic–geometric means inequality**, we have

$$\frac{9}{4} \geq \sin^2 A + \sin^2 B + \sin^2 C \geq 3\sqrt[3]{\sin^2 A \sin^2 B \sin^2 C},$$

from which (b) follows.

From $(a-b)^2+(b-c)^2+(c-a)^2 \geq 0$ or by application of **Cauchy–Schwarz inequality**, we can show that $3\left(a^2+b^2+c^2\right) \geq (a+b+c)^2$. By (e) and by setting $a = \sin A$, $b = \sin B$, $c = \sin C$, we obtain (c).

Part (f) follows from (e) and $\cos 2x = 2\cos^2 x - 1$. Finally, (g) follows from (b) and the identity

$$\sin 2A + \sin 2B + \sin 2C = 4 \sin A \sin B \sin C$$

proved in Problem 25(e).

29. Prove that

$$\frac{\tan 3x}{\tan x} = \tan\left(\frac{\pi}{3} - x\right)\tan\left(\frac{\pi}{3} + x\right)$$

for all $x \neq \frac{k\pi}{6}$, where k is in $\mathbb{Z}$.

Solution: From the **triple-angle formulas**, we have

$$\begin{aligned}\tan 3x &= \frac{3\tan x - \tan^3 x}{1 - 3\tan^2 x}\\ &= \tan x \cdot \frac{(\sqrt{3} - \tan x)(\sqrt{3} + \tan x)}{(1 - \sqrt{3}\tan x)(1 + \sqrt{3}\tan x)}\\ &= \frac{\sqrt{3} - \tan x}{1 + \sqrt{3}\tan x} \cdot \tan x \cdot \frac{\sqrt{3} + \tan x}{1 - \sqrt{3}\tan x}\\ &= \tan\left(\frac{\pi}{3} - x\right)\tan x \tan\left(\frac{\pi}{3} + x\right)\end{aligned}$$

for all $x \neq \frac{k\pi}{6}$, where k is in $\mathbb{Z}$.

30. [AMC12P 2002] Given that

$$(1 + \tan 1^\circ)(1 + \tan 2^\circ)\cdots(1 + \tan 45^\circ) = 2^n,$$

find n.

First Solution: Note that

$$\begin{aligned}1 + \tan k^\circ = 1 + \frac{\sin k^\circ}{\cos k^\circ} &= \frac{\cos k^\circ + \sin k^\circ}{\cos k^\circ}\\ &= \frac{\sqrt{2}\sin(45 + k)^\circ}{\cos k^\circ} = \frac{\sqrt{2}\cos(45 - k)^\circ}{\cos k^\circ}.\end{aligned}$$

Hence

$$(1 + \tan k^\circ)(1 + \tan(45 - k)^\circ) = \frac{\sqrt{2}\cos(45 - k)^\circ}{\cos k^\circ} \cdot \frac{\sqrt{2}\cos(k)^\circ}{\cos(45 - k)^\circ} = 2.$$

It follows that

$$\begin{aligned}&(1 + \tan 1^\circ)(1 + \tan 2^\circ)\cdots(1 + \tan 45^\circ)\\ &\quad = (1 + \tan 1^\circ)(1 + \tan 44^\circ)(1 + \tan 2^\circ)(1 + \tan 43^\circ)\\ &\qquad \cdots(1 + \tan 22^\circ)(1 + \tan 23^\circ)(1 + \tan 45^\circ)\\ &\quad = 2^{23},\end{aligned}$$

implying that $n = 23$.

Second Solution: Note that

$$\begin{aligned}
&(1 + \tan k^\circ)(1 + \tan(45 - k)^\circ) \\
&\quad = 1 + [\tan k^\circ + \tan(45 - k)^\circ] + \tan k^\circ \tan(45 - k)^\circ \\
&\quad = 1 + \tan 45^\circ[1 - \tan k^\circ \tan(45 - k)^\circ] + \tan k^\circ \tan(45 - k)^\circ \\
&\quad = 2.
\end{aligned}$$

Hence

$$\begin{aligned}
&(1 + \tan 1^\circ)(1 + \tan 2^\circ) \cdots (1 + \tan 45^\circ) \\
&\quad = (1 + \tan 1^\circ)(1 + \tan 44^\circ)(1 + \tan 2^\circ)(1 + \tan 43^\circ) \\
&\qquad \cdots (1 + \tan 22^\circ)(1 + \tan 23^\circ)(1 + \tan 45^\circ) \\
&\quad = 2^{23},
\end{aligned}$$

implying that $n = 23$.

31. [AIME 2003] Let $A = (0, 0)$ and $B = (b, 2)$ be points in the coordinate plane. Let $ABCDEF$ be a convex equilateral hexagon such that $\angle FAB = 120^\circ$, $AB \parallel DE$, $BC \parallel EF$, and $CD \parallel FA$, and the y coordinates of its vertices are distinct elements of the set $\{0, 2, 4, 6, 8, 10\}$. The area of the hexagon can be written in the form $m\sqrt{n}$, where m and n are positive integers and n is not divisible by the square of any prime. Find $m + n$.

Note: Without loss of generality, we assume that $b > 0$. (Otherwise, we can reflect the hexagon across the y axis.) Let the x coordinates of C, D, E, and F be c, d, e, and f, respectively. Note that the y coordinate of C is not 4, since if it were, the fact $|AB| = |BC|$ would imply that A, B, and C are collinear or that $c = 0$, implying that $ABCDEF$ is concave. Therefore, $F = (f, 4)$. Since $\overrightarrow{AF} = \overrightarrow{CD}$, $C = (c, 6)$ and $D = (d, 10)$, and so $E = (e, 8)$. Because the y coordinates of B, C, and D are 2, 6, and 10, respectively, and $|BC| = |CD|$, we conclude that $b = d$. Since $\overrightarrow{AB} = \overrightarrow{ED}$, $e = 0$. Let a denote the side length of the hexagon. Then $f < 0$. We need to compute

$$\begin{aligned}
[ABCDEF] &= [ABDE] + [AEF] + [BCD] = [ABDE] + 2[AEF] \\
&= b \cdot AE + (-f) \cdot AE = 8(b - f).
\end{aligned}$$

Solution:

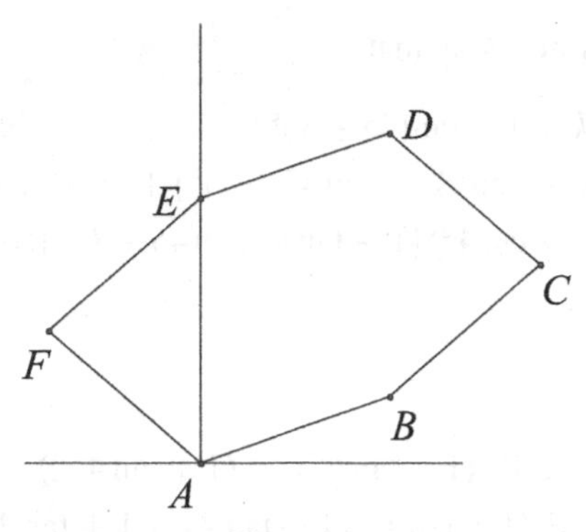

Figure 4.4.

First Solution: Note that $f^2 + 16 = |AF|^2 = a^2 = |AB|^2 = b^2 + 4$. Apply the **law of cosines** in triangle ABF to obtain $3a^2 = |BF|^2 = (b - f)^2 + 4$. We have three independent equations in three variables. Hence we can solve this system of equations. The quickest way is to note that

$$b^2 + f^2 - 2bf + 4 = (b - f)^2 + 4 = 3a^2 = a^2 + b^2 + 4 + f^2 + 16,$$

implying that $a^2 + 16 = -2bf$. Squaring both sides gives

$$a^4 + 32a^2 + 16^2 = 4b^2 f^2 = 4(a^2 - 4)(a^2 - 16) = 4a^4 - 80a^2 + 16^2,$$

or $3a^4 - 112a^2 = 0$. Hence $a^2 = \frac{112}{3}$, and so $b = \frac{10}{\sqrt{3}}$ and $f = -\frac{8}{\sqrt{3}}$. Therefore, $[ABCDEF] = 8(b - f) = 48\sqrt{3}$, and the answer to the problem is 51.

Second Solution: Let α denote the measure (in degrees) of the standard angle formed by the line AB and and the x axis. Then the standard angle formed by the line AF and the x axis is $\beta = 120^\circ + \alpha$. By considering the y coordinates of B and F, we have $a \sin \alpha = 2$ and

$$4 = a \sin(120^\circ + \alpha) = \frac{a\sqrt{3}\cos\alpha}{2} - \frac{a \sin\alpha}{2} = \frac{a\sqrt{3}\cos\alpha}{2} - 1,$$

by the **addition and subtraction formulas**. Hence $a \cos \alpha = \frac{10}{\sqrt{3}}$. Thus, by considering the x coordinates of B and F, we have $b = a \cos \alpha = \frac{10}{\sqrt{3}}$ and

$$f = a \cos(120^\circ + \alpha) = -\frac{a \cos\alpha}{2} - \frac{a\sqrt{3}\sin\alpha}{2} = -\frac{8}{\sqrt{3}}.$$

It follows that $[ABCDEF] = 48\sqrt{3}$.

Note: The vertices of the hexagon are $A = (0, 0)$, $B = \left(\frac{10}{\sqrt{3}}, 2\right)$, $C = \left(6\sqrt{3}, 6\right)$, $D = \left(\frac{10}{\sqrt{3}}, 10\right)$, $E = (0, 8)$, and $F = \left(-\frac{8}{\sqrt{3}}, 4\right)$.

32. Show that one can use a composition of trigonometry buttons such as, sin, cos, tan, $\sin^{-1}$, $\cos^{-1}$, and $\tan^{-1}$, to replace the broken reciprocal button on a calculator.

Solution: Because $\cos^{-1}\sin\theta = \pi/2 - \theta$, and $\tan(\pi/2 - \theta) = 1/\tan\theta$ for $0 < \theta < \pi/2$, we have for any $x > 0$,

$$\tan\cos^{-1}\sin\tan^{-1}x = \tan\left(\frac{\pi}{2} - \tan^{-1}x\right) = \frac{1}{x},$$

as desired. It is not difficult to check that $\tan\sin^{-1}\cos\tan^{-1}$ will also do the trick.

33. In triangle ABC, $A - B = 120°$ and $R = 8r$. Find $\cos C$.

Solution: From Problem 25(d), it follows that

$$2\sin\frac{A}{2}\sin\frac{B}{2}\sin\frac{C}{2} = \frac{1}{16},$$

or

$$\left(\cos\frac{A-B}{2} - \cos\frac{A+B}{2}\right)\sin\frac{C}{2} = \frac{1}{16},$$

by the **product-to-sum formulas**. Taking into account that $A - B = 120°$, we obtain

$$\left(\frac{1}{2} - \sin\frac{C}{2}\right)\sin\frac{C}{2} = \frac{1}{16},$$

or

$$\left(\frac{1}{4} - \sin\frac{C}{2}\right)^2 = 0,$$

yielding $\sin\frac{C}{2} = \frac{1}{4}$. Hence $\cos C = 1 - 2\sin^2\frac{C}{2} = \frac{7}{8}$.

34. Prove that in a triangle ABC,

$$\frac{a-b}{a+b} = \tan\frac{A-B}{2}\tan\frac{C}{2}.$$

Solution: From the **law of sines** and the **sum-to-product formulas**, we have

$$\frac{a-b}{a+b} = \frac{\sin A - \sin B}{\sin A + \sin B} = \frac{2\sin\frac{A-B}{2}\cos\frac{A+B}{2}}{2\sin\frac{A+B}{2}\cos\frac{A-B}{2}}$$
$$= \tan\frac{A-B}{2}\cot\frac{A+B}{2} = \tan\frac{A-B}{2}\tan\frac{C}{2},$$

as desired.

35. In triangle ABC, $\frac{a}{b} = 2+\sqrt{3}$ and $\angle C = 60°$. Find the measure of angles A and B.

Solution: From the previous problem we deduce that

$$\frac{\frac{a}{b}-1}{\frac{a}{b}+1} = \tan\frac{A-B}{2}\tan\frac{C}{2}.$$

It follows that

$$\frac{1+\sqrt{3}}{\sqrt{3}+3} = \tan\frac{A-B}{2}\cdot\frac{1}{\sqrt{3}},$$

and so $\tan\frac{A-B}{2} = 1$. Thus $A-B = 90°$, and since $A+B = 180° - C = 120°$, we obtain $A = 105°$ and $B = 15°$.

36. Let a, b, c be real numbers, all different from -1 and 1, such that $a+b+c = abc$. Prove that

$$\frac{a}{1-a^2} + \frac{b}{1-b^2} + \frac{c}{1-c^2} = \frac{4abc}{(1-a^2)(1-b^2)(1-c^2)}.$$

Solution: Let $a = \tan x$, $b = \tan y$, $c = \tan z$, where $x, y, z \neq \frac{k\pi}{4}$, for all integers k. The condition $a+b+c = abc$ translates to $\tan(x+y+z) = 0$, as indicated in notes after Problem 20(a). From the **double-angle formulas**, it follows that

$$\tan(2x+2y+2z) = \frac{2\tan(x+y+z)}{1-\tan^2(x+y+z)} = 0.$$

Hence

$$\tan 2x + \tan 2y + \tan 2z = \tan 2x \tan 2y \tan 2z,$$

using a similar argument to the one in Problem 20(a). This implies that

$$\frac{2\tan x}{1-\tan^2 x} + \frac{2\tan y}{1-\tan^2 y} + \frac{2\tan z}{1-\tan^2 z} = \frac{2\tan x}{1-\tan^2 x} \cdot \frac{2\tan y}{1-\tan^2 y} \cdot \frac{2\tan z}{1-\tan^2 z},$$

and the conclusion follows.

37. Prove that a triangle ABC is isosceles if and only if

$$a\cos B + b\cos C + c\cos A = \frac{a+b+c}{2}.$$

Solution: By the **extended law of sines,** $a = 2R\sin A$, $b = 2R\sin B$, and $c = 2R\sin C$. The desired identity is equivalent to

$$2\sin A\cos B + 2\sin B\cos C + 2\sin C\cos A = \sin A + \sin B + \sin C,$$

or

$$\begin{aligned}\sin(A+B) + \sin(A-B) + \sin(B+C)\\ + \sin(B-C) + \sin(C+A) + \sin(C-A)\\ = \sin A + \sin B + \sin C.\end{aligned}$$

Because $A + B + C = 180°$, $\sin(A+B) = \sin C$, $\sin(B+C) = \sin A$, $\sin(C+A) = \sin B$. The last equality simplifies to

$$\sin(A-B) + \sin(B-C) + \sin(C-A) = 0,$$

which in turn is equivalent to

$$4\sin\frac{A-B}{2}\sin\frac{B-C}{2}\sin\frac{C-A}{2} = 0,$$

by Problem 15. The conclusion now follows.

38. Evaluate

$$\cos a \cos 2a \cos 3a \cdots \cos 999a,$$

where $a = \frac{2\pi}{1999}$.

Solution: Let P denote the desired product, and let

$$Q = \sin a \sin 2a \sin 3a \cdots \sin 999a.$$

Then

$$\begin{aligned}2^{999} PQ &= (2\sin a\cos a)(2\sin 2a\cos 2a)\cdots(2\sin 999a\cos 999a)\\ &= \sin 2a \sin 4a \cdots \sin 1998a\\ &= (\sin 2a\sin 4a\cdots\sin 998a)[-\sin(2\pi - 1000a)]\\ &\quad\cdot[-\sin(2\pi - 1002a)]\cdots[-\sin(2\pi - 1998a)]\\ &= \sin 2a\sin 4a\cdots\sin 998a\sin 999a\sin 997a\cdots\sin a = Q.\end{aligned}$$

It is easy to see that $Q \neq 0$. Hence the desired product is $P = \frac{1}{2^{999}}$.

39. Determine the minimum value of

$$\frac{\sec^4\alpha}{\tan^2\beta} + \frac{\sec^4\beta}{\tan^2\alpha}$$

over all $\alpha, \beta \neq \frac{k\pi}{2}$, where k is in $\mathbb{Z}$.

Solution: Set $a = \tan^2\alpha$ and $b = \tan^2\beta$. It suffices to determine the minimum value of

$$\frac{(a+1)^2}{b} + \frac{(b+1)^2}{a},$$

with $a, b \geq 0$. We have

$$\begin{aligned}\frac{(a+1)^2}{b} + \frac{(b+1)^2}{a} &= \frac{a^2+2a+1}{b} + \frac{b^2+2b+1}{a}\\ &= \left(\frac{a^2}{b} + \frac{1}{b} + \frac{b^2}{a} + \frac{1}{a}\right) + 2\left(\frac{a}{b} + \frac{b}{a}\right)\\ &\geq 4\sqrt[4]{\frac{a^2}{b}\cdot\frac{1}{b}\cdot\frac{b^2}{a}\cdot\frac{1}{a}} + 4\sqrt{\frac{a}{b}\cdot\frac{b}{a}} = 8,\end{aligned}$$

by the **arithmetic–geometric means inequality**. Equality holds when $a = b = 1$; that is, $\alpha = \pm 45^\circ + k\cdot 180^\circ$ and $\beta = \pm 45^\circ + k \cdot 180^\circ$, for integers k.

40. Find all pairs (x, y) of real numbers with $0 < x < \frac{\pi}{2}$ such that

$$\frac{(\sin x)^{2y}}{(\cos x)^{y^2/2}} + \frac{(\cos x)^{2y}}{(\sin x)^{y^2/2}} = \sin 2x.$$

Solution: The **arithmetic–geometric means inequality** gives

$$\frac{(\sin x)^{2y}}{(\cos x)^{y^2/2}} + \frac{(\cos x)^{2y}}{(\sin x)^{y^2/2}} \geq 2(\sin x \cos x)^{y-y^2/4}.$$

It follows that

$$2 \sin x \cos x = \sin 2x \geq 2(\sin x \cos x)^{y-y^2/4},$$

and because $\sin x \cos x < 1$, it follows that $1 \leq y - y^2/4$, or $(1 - y/2)^2 \leq 0$. It follows that all the equalities hold; that is, $y = 2$ and $\sin x = \cos x$, and so there is a unique solution: $(x, y) = \left(\frac{\pi}{4}, 2\right)$.

41. Prove that $\cos 1^\circ$ is an irrational number.

Solution: Assume, for the sake of contradiction, that $\cos 1^\circ$ is rational. Then so is $\cos 2^\circ = 2\cos^2 1^\circ - 1$. Using the identity

$$\cos(n^\circ + 1^\circ) + \cos(n^\circ - 1^\circ) = 2 \cos n^\circ \cos 1^\circ, \qquad (*)$$

we obtain by strong induction that $\cos n^\circ$ is rational for all integers $n \geq 1$. But this is clearly false, because, for example, $\cos 30^\circ$ is not rational, yielding a contradiction.

Note: For the reader not familiar with the idea of induction. We can reason in the following way. Under the assumption that both $\cos 1^\circ$ and $\cos 2^\circ$ are rational, relation $(*)$ implies that $\cos 3^\circ$ is rational, by setting $n = 2$ in the relation $(*)$. Similarly, by the assumption that both $\cos 2^\circ$ and $\cos 3^\circ$ are rational, relation $(*)$ implies that $\cos 4^\circ$ is rational, by setting $n = 4$ in the relation $(*)$. And so on. We conclude that $\cos n^\circ$ is rational, for all positive integers n, under the assumption that $\cos 1^\circ$ is rational.

42. [USAMO 2002 proposal, by Cecil Rousseau] Find the maximum value of

$$S = (1 - x_1)(1 - y_1) + (1 - x_2)(1 - y_2)$$

if $x_1^2 + x_2^2 = y_1^2 + y_2^2 = c^2$.

Solution: If we interpret x_1 and x_2 are the coordinates of a point; that is, assume that $P = (x_1, x_2)$, then P lies on a circle centered at the origin with radius c. We can describe the circle parametrically; that is, write $x_1 = c\cos\theta$, $x_2 = c\sin\theta$, and similarly, $y_1 = c\cos\phi$, $y_2 = c\sin\phi$. Then

$$\begin{aligned} S &= 2 - c(\cos\theta + \sin\theta + \cos\phi + \sin\phi) + c^2(\cos\theta\cos\phi + \sin\theta\sin\phi) \\ &= 2 - \sqrt{2}c[\sin(\theta + \pi/4) + \sin(\phi + \pi/4)] + c^2\cos(\theta - \phi) \\ &\le 2 + 2\sqrt{2}c + c^2 = (\sqrt{2} + c)^2, \end{aligned}$$

with equality at $\theta = \phi = 5\pi/4$, that is, $x_1 = x_2 = y_1 = y_2 = \frac{-c\sqrt{2}}{2}$.

43. Prove that

$$\frac{\sin^3 a}{\sin b} + \frac{\cos^3 a}{\cos b} \ge \sec(a - b)$$

for all $0 < a, b < \frac{\pi}{2}$.

Solution: Multiplying the two sides of the inequality by $\sin a\sin b + \cos a\cos b = \cos(a - b)$, we obtain the equivalent form

$$\left(\frac{\sin^3 a}{\sin b} + \frac{\cos^3 a}{\cos b}\right)(\sin a\sin b + \cos a\cos b) \ge 1.$$

But this follows from **Cauchy–Schwarz inequality** because according to this inequality, the left-hand side is greater than or equal to $(\sin^2 a + \cos^2 a)^2 = 1$.

44. If $\sin\alpha\cos\beta = -\frac{1}{2}$, what are the possible values of $\cos\alpha\sin\beta$?

Solution: Note that

$$\sin(\alpha + \beta) = \sin\alpha\cos\beta + \cos\alpha\sin\beta = -\frac{1}{2} + \cos\alpha\sin\beta.$$

Because $-1 < \sin(\alpha + \beta) \le 1$, it follows that $-\frac{1}{2} \le \cos\alpha\sin\beta < \frac{3}{2}$. Similarly, because $\sin(\alpha - \beta) = \sin\alpha\cos\beta - \cos\alpha\sin\beta$, we conclude that $-\frac{3}{2} \le \cos\alpha\sin\beta < \frac{1}{2}$. Combining the above results shows that

$$-\frac{1}{2} \le \cos\alpha\sin\beta \le \frac{1}{2}.$$

But we have not shown that indeed, $\cos\alpha\sin\beta$ can obtain all values in the interval $\left[-\frac{1}{2}, \frac{1}{2}\right]$. To do this, we consider

$$\begin{aligned}(\cos\alpha\sin\beta)^2 &= (1-\sin^2\alpha)(1-\cos^2\beta)\\ &= 1-(\sin^2\alpha+\cos^2\beta)+\sin^2\alpha\cos^2\beta\\ &= \frac{5}{4}-(\sin^2\alpha+\cos^2\beta)\\ &= \frac{5}{4}-(\sin\alpha+\cos\beta)^2+2\sin\alpha\cos\beta\\ &= \frac{1}{4}-(\sin\alpha+\cos\beta)^2.\end{aligned}$$

Let $x=\sin\alpha$ and $y=\cos\beta$. Then $-1\le x, y, \le 1$ and $xy=-\frac{1}{2}$. Consider the range of the sum $s=\sin\alpha+\cos\beta=x+y$. If $xy=-\frac{1}{2}$ and $x+y=s$, then x and y are the roots of the quadratic equation

$$u^2-su-\frac{1}{2}=0. \qquad (*)$$

Thus, $\{x, y\}=\left\{\frac{s+\sqrt{s^2+2}}{2}, \frac{s-\sqrt{s^2+2}}{2}\right\}$. By checking the boundary condition $\frac{s+\sqrt{s^2+2}}{2}\le 1$, we obtain $s\le\frac{1}{2}$. By checking similar boundary conditions, we conclude that the equation $(*)$ has a pair of solutions x and y with $-1\le x, y\le 1$ for all $-\frac{1}{2}\le s\le\frac{1}{2}$. Because both the sine and cosine functions are surjective functions from $\mathbb{R}$ to the interval $[-1, 1]$, the range of $s=\sin\alpha+\cos\beta$ is $\left[-\frac{1}{2}, \frac{1}{2}\right]$ for $\sin\alpha\cos\beta=-\frac{1}{2}$. Thus, the range of s^2 is $\left[0, \frac{1}{2}\right]$. Thus the range of $(\cos\alpha\sin\beta)^2$ is $\left[0, \frac{1}{4}\right]$, and so the range of $\cos\alpha\sin\beta$ is $\left[-\frac{1}{2}, \frac{1}{2}\right]$.

45. Let a, b, c be real numbers. Prove that

$$(ab+bc+ca-1)^2\le(a^2+1)(b^2+1)(c^2+1).$$

Solution: Let $a=\tan x$, $b=\tan y$, $c=\tan z$ with $-\frac{\pi}{2}<x, y, z<\frac{\pi}{2}$. Then $a^2+1=\sec^2 x$, $b^2+1=\sec^2 y$, and $c^2+1=\sec^2 z$. Multiplying by $\cos^2 x\cos^2 y\cos^2 z$ on both sides of the desired inequality gives

$$[(ab+bc+ca-1)\cos x\cos y\cos z]^2\le 1.$$

Note that

$$\begin{aligned}(ab+bc)\cos x\cos y\cos z &= \sin x\sin y\cos z+\sin y\sin z\cos x\\ &= \sin y\sin(x+z)\end{aligned}$$

and

$$(ca - 1)\cos x \cos y \cos z = \sin z \sin x \cos y - \cos x \cos y \cos z$$
$$= -\cos y \cos(x + z).$$

Consequently, we obtain

$$[(ab + bc + ca - 1)\cos x \cos y \cos z]^2$$
$$= [\sin y \sin(x + z) - \cos y \cos(x + z)]^2$$
$$= \cos^2(x + y + z) \leq 1,$$

as desired.

46. Prove that

$$(\sin x + a\cos x)(\sin x + b\cos x) \leq 1 + \left(\frac{a+b}{2}\right)^2.$$

Solution: If $\cos x = 0$, the desired inequality reduces to $\sin^2 x \leq 1 + \left(\frac{a+b}{2}\right)^2$, which is clearly true. We assume that $\cos x \neq 0$. Dividing both sides of the desired inequality by $\cos^2 x$ gives

$$(\tan x + a)(\tan x + b) \leq \left[1 + \left(\frac{a+b}{2}\right)^2\right]\sec^2 x.$$

Set $t = \tan x$. Then $\sec^2 x = 1 + t^2$. The above inequality reduces to

$$t^2 + (a+b)t + ab \leq \left(\frac{a+b}{2}\right)^2 t^2 + t^2 + \left(\frac{a+b}{2}\right)^2 + 1,$$

or

$$\left(\frac{a+b}{2}\right)^2 t^2 + 1 - (a+b)t + \left(\frac{a+b}{2}\right)^2 - ab \geq 0.$$

The last inequality is equivalent to

$$\left(\frac{(a+b)t}{2} - 1\right)^2 + \left(\frac{a-b}{2}\right)^2 \geq 0,$$

which is evident.

47. Prove that

$$|\sin a_1| + |\sin a_2| + \cdots + |\sin a_n| + |\cos(a_1 + a_2 + \cdots + a_n)| \geq 1.$$

Solution: We proceed by induction on n. The base case holds, because

$$|\sin a_1| + |\cos a_1| \geq \sin^2 a_1 + \cos^2 a_1 = 1.$$

For the inductive step, in order to prove that

$$|\sin a_1| + |\sin a_2| + \cdots + |\sin a_{n+1}| + |\cos(a_1 + a_2 + \cdots + a_{n+1})| \geq 1,$$

it suffices to show that

$$|\sin a_{n+1}| + |\cos(a_1 + a_2 + \cdots + a_{n+1})| \geq |\cos(a_1 + a_2 + \cdots + a_n)|$$

for all real numbers $a_1, a_2, \ldots, a_{n+1}$. Let $s_k = a_1 + a_2 + \cdots + a_k$, for $k = 1, 2, \ldots, n+1$. The last inequality becomes $|\sin a_{n+1}| + |\cos s_{n+1}| \geq |\cos s_n|$. Indeed, by the **addition and subtraction formulas**, we have

$$\begin{aligned}
|\cos s_n| &= |\cos(s_{n+1} - a_{n+1})| \\
&= |\cos s_{n+1} \cos a_{n+1} + \sin s_{n+1} \sin a_{n+1}| \\
&= |\cos s_{n+1} \cos a_{n+1}| + |\sin s_{n+1} \sin a_{n+1}| \\
&\leq |\cos s_{n+1}| + |\sin a_{n+1}|,
\end{aligned}$$

as desired.

48. [Russia 2003, by Nazar Agakhanov] Find all angles α for which the three-element set

$$S = \{\sin\alpha, \sin 2\alpha, \sin 3\alpha\}$$

is equal to the set

$$T = \{\cos\alpha, \cos 2\alpha, \cos 3\alpha\}.$$

Solution: The answers are $\alpha = \frac{\pi}{8} + \frac{k\pi}{2}$ for all integers k.

Because $S = T$, the sums of the elements in S and T are equal to each other; that is,

$$\sin\alpha + \sin 2\alpha + \sin 3\alpha = \cos\alpha + \cos 2\alpha + \cos 3\alpha.$$

Applying the **sum-to-product formulas** to the first and the third summands on each side of the last equation gives

$$2\sin 2\alpha\cos\alpha + \sin 2\alpha = 2\cos 2\alpha\cos\alpha + \cos 2\alpha,$$

or

$$\sin 2\alpha(2\cos\alpha + 1) = \cos 2\alpha(2\cos\alpha + 1).$$

If $2\cos\alpha + 1 = 0$, then $\cos\alpha = -\frac{1}{2}$, and so $\alpha = \pm\frac{2\pi}{3} + 2k\pi$ for all integers k. It is then not difficult to check that $S \neq T$ and both of S and T are not three-element sets.

It follows that $2\cos\alpha + 1 \neq 0$, implying that $\sin 2\alpha = \cos 2\alpha$; that is, $\tan 2\alpha = 1$. The possible answers are $\alpha = \frac{\pi}{8} + \frac{k\pi}{2}$ for all integers k. Because $\frac{\pi}{8} + \frac{3\pi}{8} = \frac{\pi}{2}$, $\cos\frac{\pi}{8} = \sin\frac{3\pi}{8}$. It not difficult to check that all such angles satisfy the conditions of the problem.

49. Let $\{T_n(x)\}_{n=0}^{\infty}$ be the sequence of polynomials such that $T_0(x) = 1$, $T_1(x) = x$, $T_{i+1} = 2xT_i(x) - T_{i-1}(x)$ for all positive integers i. The polynomial $T_n(x)$ is called the nth **Chebyshev polynomial**.

    (a) Prove that $T_{2n+1}(x)$ and $T_{2n}(x)$ are odd and even functions, respectively;

    (b) Prove that $T_{n+1}(x) > T_n(x) > 1$ for real numbers x with $x > 1$;

    (c) Prove that $T_n(\cos\theta) = \cos(n\theta)$ for all nonnegative integers n;

    (d) Determine all the roots of $T_n(x)$;

    (e) Determine all the roots of $P_n(x) = T_n(x) - 1$.

Solution: Parts (a) and (b) are simple facts that will be useful in establishing (e). We present them together.

(a) We apply strong induction on n. Note that $T_0 = 1$ and $T_1 = x$ are even and odd, respectively. Assume that T_{2n-1} and T_{2n} are odd and even, respectively. Then $2xT_{2n}$ is odd, and so $T_{2n+1} = 2xT_{2n} - T_{2n-1}$ is odd. Thus $2xT_{2n+1}$ is even, and so $T_{2n+2} = 2xT_{2n+1} - T_{2n}$ is even. This completes our induction.

(b) We apply strong induction on n. For $n = 0$, $T_1(x) = x > 1 = T_0(x)$ for $x > 1$. Assume that $T_{n+1}(x) > T_n(x) > 1$ for $x > 1$ and $n \le k$, where k is some nonnegative integer. For $n = k + 1$, the induction hypothesis yields

$$\begin{aligned} T_{k+2}(x) &= 2xT_{k+1}(x) - T_k(x) > 2T_{k+1}(x) - T_k(x) \\ &= T_{k+1}(x) + T_{k+1}(x) - T_k(x) > T_{k+1}(x), \end{aligned}$$

completing our induction.

(c) We again apply strong induction on n. The base cases for $n = 0$ and $n = 1$ are trivial. Assume that $T_n(\cos\theta) = \cos(n\theta)$ for $n \le k$, where k is some positive integer. The induction hypothesis gives

$$\begin{aligned} T_{k+1}(\cos\theta) &= 2\cos\theta\, T_k(\cos\theta) - T_{k-1}(\cos\theta) \\ &= 2\cos\theta\cos k\theta - \cos[(k-1)\theta]. \end{aligned}$$

By the **product-to-sum formulas**, we have

$$2\cos\theta\cos k\theta = \cos[(k+1)\theta] + \cos[(k-1)\theta].$$

It follows that $T_{k+1}(\cos\theta) = \cos[(k+1)\theta]$, completing our induction.

(d) It is clear that T_n is a polynomial of degree n, and so it has at most n real roots. Note that $y = \cos x$ is a one-to-one and onto mapping from the interval $\left[0, \frac{\pi}{2}\right]$. By (c), we conclude that T_n has exactly n distinct real roots, and they form the set

$$S = \left\{ \cos\frac{k\pi}{2n},\ k = 1, 3, \ldots, 2n-1 \right\}.$$

(e) By (a), T_n is either even or odd, and so by (b), $|T_n(x)| > 1$ for $x < -1$. Thus, all the roots of P_n lie in the interval $[-1, 1]$. We consider two cases. Assume first that n is even. A real number is a root of P_n if and only if it is in the set

$$S_e = \left\{ \cos\frac{k\pi}{n},\ k = 0, 2, \ldots, n \right\}.$$

Assume next that n is odd. A real number is a root of P_n if and only if it is in the set

$$S_e = \left\{ \cos\frac{k\pi}{n},\ k = 0, 2, \ldots, n-1 \right\}.$$

50. [Canada 1998] Let ABC be a triangle with $\angle BAC = 40°$ and $\angle ABC = 60°$. Let D and E be the points lying on the sides AC and AB, respectively, such that $\angle CBD = 40°$ and $\angle BCE = 70°$. Segments BD and CE meet at F. Show that $AF \perp BC$.

Solution:

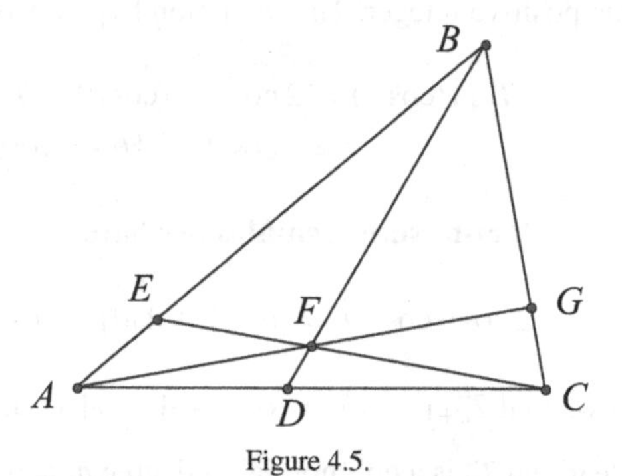

Figure 4.5.

Note that $\angle ABD = 20°$, $\angle BCA = 80°$, and $\angle ACE = 10°$. Let G be the foot of the altitude from A to BC. Then $\angle BAG = 90° - \angle ABC = 30°$ and $\angle CAG = 90° - \angle BCA = 10°$. Now,

$$\begin{aligned}\frac{\sin\angle BAG \sin\angle ACE \sin\angle CBD}{\sin\angle CAG \sin\angle BCE \sin\angle ABD} &= \frac{\sin 30° \sin 10° \sin 40°}{\sin 10° \sin 70° \sin 20°}\\ &= \frac{\frac{1}{2}(\sin 10°)(2\sin 20° \cos 20°)}{\sin 10° \cos 20° \sin 20°}\\ &= 1.\end{aligned}$$

Then by the trigonometric form of **Ceva's theorem**, lines AG, BD, and CE are concurrent. Therefore, F lies on segment AG, and so line AF is perpendicular to the line BC, as desired.

51. [IMO 1991] Let S be an interior point of triangle ABC. Show that at least one of $\angle SAB$, $\angle SBC$, and $\angle SCA$ is less than or equal to $30°$.

First Solution: The given conditions in the problem motivate us to consider the **Brocard point** P of triangle ABC with $\alpha = \angle PAB = \angle PBC = \angle PCA$. Because S (see Figure 4.6) lies inside or on the boundary of at least one of the triangles PAB, PBC, and PCA, at least one of $\angle SAB$, $\angle SBC$, and $\angle SCA$ is less than or equal to α. It suffices to show that $\alpha \le 30°$; that is, $\sin\alpha \le \frac{1}{2}$ or $\csc^2\alpha \ge 4$, by considering the range of α.

We have shown that

$$\csc^2\alpha = \csc^2 A + \csc^2 B + \csc^2 C.$$

By Problem 28(e) and **Cauchy–Schwarz inequality**, we have

$$\frac{9}{4}\csc^2\alpha \geq \left(\sin^2 A + \sin^2 B + \sin^2 C\right)\left(\csc^2 A + \csc^2 B + \csc^2 C\right) \geq 9,$$

implying that $\csc^2\alpha \geq 4$, as desired.

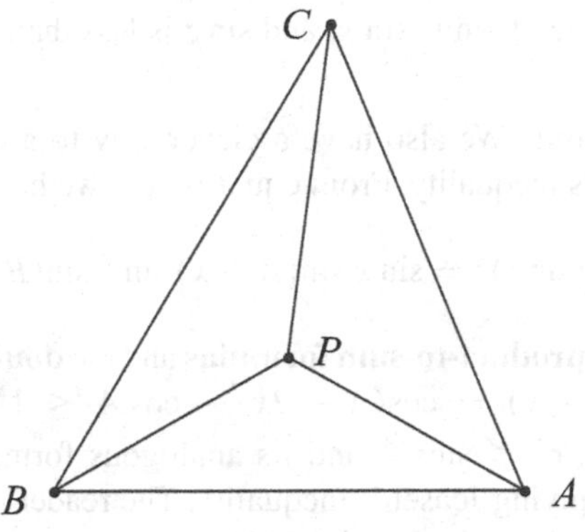

Figure 4.6.

Second Solution: We use radian measure in this solution; that is, we want to show that one of $\angle SAB$, $\angle SBC$, and $\angle SCA$ is less than or equal to $\frac{\pi}{6}$. Set $x = \angle SAB$, $y = \angle SBC$, and $z = \angle SCA$. Let d_a, d_b and d_c denote the distance from S to sides BC, CA, and AB. Then

$$\begin{aligned} d_c &= SA\sin x = SB\sin(B-y), \\ d_a &= SB\sin y = SC\sin(C-z), \\ d_b &= SC\sin z = SA\sin(A-x). \end{aligned}$$

Multiplying the last three equations together gives

$$\sin x \sin y \sin z = \sin(A-x)\sin(B-y)\sin(C-z). \qquad (*)$$

If $x+y+z \leq \frac{\pi}{2}$, then the conclusion of the problem is clearly true. Now we assume that $x+y+z > \frac{\pi}{2}$; that is, $(A-x)+(B-y)+(C-z) < \frac{\pi}{2}$. Now we consider the function $f(x) = \ln(\sin x)$, where $0 < x < \frac{\pi}{2}$. Then the first derivative of $f(x)$ is $f'(x) = \frac{\cos x}{\sin x} = \cot x$, and the second derivative is $f''(x) = -\csc^2 x < 0$. Hence $f(x)$ is **concave down**. By **Jensen's inequality**, we have

$$\begin{aligned} &\frac{1}{3}(\ln\,\sin(A-x) + \ln\,\sin(B-y) + \ln\,\sin(C-z)) \\ &\leq \ln\sin\frac{(A-x)+(B-y)+(C-z)}{3}, \end{aligned}$$

implying that

$$\ln\left(\sin(A-x)\sin(B-y)\sin(C-z)\right)^{\frac{1}{3}} \le \ln\sin\frac{6}{\pi} = \ln\frac{1}{2},$$

or $\sin(A-x)\sin(B-y)\sin(C-z) \le \frac{1}{8}$. Thus $\sin x \sin y \sin z \le \frac{1}{8}$, implying that at least one of $\sin x$, $\sin y$, and $\sin z$ is less than or equal to $\frac{1}{2}$, as desired.

Third Solution: We also have a clever way to apply equation $(*)$ without using Jensen's inequality. From equation $(*)$ we have

$$(\sin x \sin y \sin z)^2 = \sin x \sin(A-x) \sin y \sin(B-y) \sin z \sin(C-z).$$

Applying the **product-to-sum formulas** and the **double-angle formulas** gives $2\sin x \sin(A-x) = \cos(A-2x) - \cos A \le 1 - \cos A = 2\sin^2\frac{A}{2}$, or $\sin x \sin(A-x) \le \sin^2\frac{A}{2}$ and its analogous forms. (This step can also be carried by applying Jensen's inequality. The reader might want to do so as an exercise.) It follows, by Problem 23(a), that

$$\sin x \sin y \sin z \le \sin\frac{A}{2}\sin\frac{B}{2}\sin\frac{C}{2} \le \frac{1}{8},$$

from which our desired result follows.

52. Let $a = \frac{\pi}{7}$.

(a) Show that $\sin^2 3a - \sin^2 a = \sin 2a \sin 3a$;

(b) Show that $\csc a = \csc 2a + \csc 4a$;

(c) Evaluate $\cos a - \cos 2a + \cos 3a$;

(d) Prove that $\cos a$ is a root of the equation $8x^3 + 4x^2 - 4x - 1 = 0$;

(e) Prove that $\cos a$ is irrational;

(f) Evaluate $\tan a \tan 2a \tan 3a$;

(g) Evaluate $\tan^2 a + \tan^2 2a + \tan^2 3a$;

(h) Evaluate $\tan^2 a \tan^2 2a + \tan^2 2a \tan^2 3a + \tan^2 3a \tan^2 a$.

(i) Evaluate $\cot^2 a + \cot^2 2a + \cot^2 3a$.

Solution: Many of the desired results are closely related. Parts (d) and (e) will be presented together, as will (f), (g), (h), and (i).

(a) By the **sum-to-product**, the **difference-to-product** and the **double-angle formulas**, we have

$$\begin{aligned}\sin^2 3a - \sin^2 a &= (\sin 3a + \sin a)(\sin 3a - \sin a)\\ &= (2\sin 2a\cos a)(2\sin a\cos 2a)\\ &= (2\sin 2a\cos 2a)(2\sin a\cos a)\\ &= \sin 4a\sin 2a = \sin 2a\sin 3a,\end{aligned}$$

as desired. The last identity is evident by noting that $4a + 3a = \pi$ (and so $\sin 3a = \sin 4a$).

(b) It suffices to show that

$$\sin 2a\sin 4a = \sin a(\sin 2a + \sin 4a),$$

or

$$2\sin a\cos a\sin 4a = \sin a(2\sin 3a\cos a),$$

by the sum-to-product formulas.

(c) The answer is $\frac{1}{2}$. It suffices to show that $\cos 2a + \cos 4a + \cos 6a = -\frac{1}{2}$. This is a special case ($n = 3$) of a more general result:

$$t = \cos 2x + \cos 4x + \cdots + \cos 2nx = -\frac{1}{2},$$

where $x = \frac{\pi}{2n+1}$. Indeed, applying the **product-to-sum formulas** gives $2\sin x\cos kx = \sin(k+1)x - \sin(k-1)x$, and so

$$\begin{aligned}2t\sin x &= 2\sin x(\cos 2x + \cos 4x + \cdots + \cos 2nx)\\ &= [\sin 3x - \sin x] + [\sin 5x - \sin 3x]\\ &\quad + \cdots + [\sin(2n+1)x - \sin(2n-1)x]\\ &= \sin(2n+1)x - \sin x = -\sin x,\end{aligned}$$

from which the desired equality follows.

(d) Because $3a + 4a = \pi$, it follows that $\sin 3a = \sin 4a$. The double-angle and **triple-angle formulas** yield

$$\sin a(3 - 4\sin^2 a) = 2\sin 2a\cos 2a = 4\sin a\cos a\cos 2a,$$

or $3 - 4(1 - \cos^2 a) = 4\cos a(2\cos^2 a - 1)$. It follows that

$$8\cos^3 a - 4\cos^2 a - 4\cos a + 1 = 0,$$

establishing (c). Thus $u = 2\cos a$ is the root of the cubic equation

$$u^3 - u^2 - 2u + 1 = 0. \qquad (*)$$

By **Gauss's lemma**, the only possible rational roots of the above cubic equation are 1 and -1. It is easy to see that neither is a root. Hence the above equation has no rational root, implying that $2\cos a$ is not rational. Therefore, $\cos a$ is not rational.

Note: Although converting to equation $(*)$ is not necessary, it is a very effective technique. Instead of checking of eight possible rational roots from the set $\left\{\pm\frac{1}{8}, \pm\frac{1}{4}, \pm\frac{1}{2}, \pm 1\right\}$ of the equation

$$8x^3 - 4x^2 - 4x + 1 = 0,$$

we need to check only two possibilities for equation $(*)$.

(f) Because $3a + 4a = \pi$, it follows that $\tan 3a + \tan 4a = 0$. The double-angle and the addition and subtraction formulas yield

$$\frac{\tan a + \tan 2a}{1 - \tan a \tan 2a} + \frac{2\tan 2a}{1 - \tan^2 2a} = 0,$$

or

$$\tan a + 3\tan 2a - 3\tan a \tan^2 2a - \tan^3 2a = 0.$$

Set $\tan a = x$. Then $\tan 2a = \frac{2\tan a}{1-\tan^2 a} = \frac{2x}{1-x^2}$. Hence

$$x + \frac{6x}{1-x^2} - \frac{12x^3}{(1-x^2)^2} - \frac{8x^3}{(1-x^2)^3} = 0,$$

or

$$\left(1-x^2\right)^3 + 6\left(1-x^2\right)^2 - 12x^2\left(1-x^2\right) - 8x^2 = 0.$$

Expanding the left-hand side of the above equation gives

$$x^6 - 21x^4 + 35x^2 - 7 = 0. \qquad (\dagger)$$

Thus $\tan a$ is a root of the above equation. Note that $6a + 8a = 2\pi$ and $9a + 12a = 3\pi$, and so $\tan[3(2a)] + \tan[4(2a)] = 0$ and $\tan[3(3a)] + \tan[4(3a)] = 0$. Hence $\tan 2a$ and $\tan 3a$ are the also the roots of equation $(\dagger)$. Therefore, $\tan^2 ka$, $k = 1, 2, 3$, are the distinct roots of the cubic equation

$$x^3 - 21x^2 + 35x - 7 = 0.$$

By **Viète's theorem**, we have

$$\tan^2 a + \tan^2 2a + \tan^2 3a = 21;$$
$$\tan^2 a \tan^2 2a + \tan^2 2a \tan^2 3a + \tan^2 3a \tan^2 a = 35;$$
$$\tan^2 a \tan^2 2a \tan^2 3a = 7.$$

Thus the answers for (f), (g), (h), and (i) are $\sqrt{7}$, 21, 35, and 5, respectively.

Note: It is not difficult to check that the roots of the equation (†) are $\tan\frac{\pi}{7}$, $\tan\frac{2\pi}{7}, \ldots, \tan\frac{6\pi}{7}$. On the other hand, it is interesting to note that $1, -21, 35, -7$, the coefficients of equation (†) are $\binom{7}{0}, -\binom{7}{2}, \binom{7}{4}, -\binom{7}{6}$. In general, we have the following result: For positive integers n, let $a_n = \frac{\pi}{2n+1}$. Then $\sin(2n+1)a_n = 0$. The **expansion formulas** give

$$\begin{aligned}
0 &= \sin(2n+1)a_n \\
&= \binom{2n+1}{1}\cos^{2n} a_n \sin a_n - \binom{2n+1}{3}\cos^{2n-2} a_n \sin^3 a_n \\
&\quad + \binom{2n+1}{5}\cos^{2n-4} a_n \sin^5 a_n - \cdots \\
&= \cos^{2n+1} a_n \left[\binom{2n+1}{1}\tan a_n - \binom{2n+1}{3}\tan^3 a_n \right. \\
&\qquad\qquad \left. + \binom{2n+1}{5}\tan^5 a_n - \cdots\right].
\end{aligned}$$

Because $\cos a_n \neq 0$, it follows that

$$\binom{2n+1}{1}\tan a_n - \binom{2n+1}{3}\tan^3 a_n + \binom{2n+1}{5}\tan^5 a_n - \cdots = 0;$$

that is, $\tan a_n$ is a root of the equation

$$\binom{2n+1}{1}x - \binom{2n+1}{3}x^3 + \cdots + (-1)^n\binom{2n+1}{2n+1}x^{2n+1} = 0,$$

or

$$\binom{2n+1}{0}x^{2n} - \binom{2n+1}{2}x^{2n-2} + \cdots + (-1)^n\binom{2n+1}{2n} = 0.$$

It is not difficult to see that the roots of the above equation are $\tan\frac{\pi}{2n+1}$, $\tan\frac{2\pi}{2n+1}, \ldots, \tan\frac{2n\pi}{2n+1}$. It is also not difficult to see that the roots of the equation

$$\binom{2n+1}{0}x^{n} - \binom{2n+1}{2}x^{n-1} + \cdots + (-1)^n\binom{2n+1}{2n} = 0$$

are $\tan^2\frac{\pi}{2n+1}, \tan^2\frac{2\pi}{2n+1}, \ldots, \tan^2\frac{n\pi}{2n+1}$. By Viète's theorem, we can obtain more general results, such as

$$\sum_{k=1}^{n}\cot^2\frac{k\pi}{2n+1} = \frac{\binom{2n+1}{2n-2}}{\binom{2n+1}{2n}} = \frac{n(2n-1)}{3}.$$

5
Solutions to Advanced Problems

1. Two exercises on $\sin k^\circ \sin(k+1)^\circ$:

   (a) [AIME2 2000] Find the smallest positive integer n such that

$$\frac{1}{\sin 45^\circ \sin 46^\circ} + \frac{1}{\sin 47^\circ \sin 48^\circ} + \cdots + \frac{1}{\sin 133^\circ \sin 134^\circ} = \frac{1}{\sin n^\circ}.$$

   (b) Prove that

$$\frac{1}{\sin 1^\circ \sin 2^\circ} + \frac{1}{\sin 2^\circ \sin 3^\circ} + \cdots + \frac{1}{\sin 89^\circ \sin 90^\circ} = \frac{\cos 1^\circ}{\sin^2 1^\circ}.$$

Solution: Note that

$$\begin{aligned} \sin 1^\circ &= \sin[(x+1)^\circ - x^\circ] \\ &= \sin(x+1)^\circ \cos x^\circ - \cos(x+1)^\circ \sin x^\circ. \end{aligned}$$

Thus

$$\begin{aligned} \frac{\sin 1^\circ}{\sin x^\circ \sin(x+1)^\circ} &= \frac{\cos x^\circ \sin(x+1)^\circ - \sin x^\circ \cos(x+1)^\circ}{\sin x^\circ \sin(x+1)^\circ} \\ &= \cot x^\circ - \cot(x+1)^\circ. \end{aligned}$$

(a) Multiplying both sides of the given equation by $\sin 1^\circ$, we have

$$\begin{aligned}\frac{\sin 1^\circ}{\sin n^\circ} &= (\cot 45^\circ - \cot 46^\circ) + (\cot 47^\circ - \cot 48^\circ)\\ &\quad + \cdots + (\cot 133^\circ - \cot 134^\circ)\\ &= \cot 45^\circ - (\cot 46^\circ + \cot 134^\circ) + (\cot 47^\circ + \cot 133^\circ)\\ &\quad - \cdots + (\cot 89^\circ + \cot 91^\circ) - \cot 90^\circ\\ &= 1.\end{aligned}$$

Therefore, $\sin n^\circ = \sin 1^\circ$, and the least possible integer value for n is 1.

(b) The left-hand side of the desired equation is equal to

$$\begin{aligned}\sum_{k=1}^{89} \frac{1}{\sin k^\circ \sin(k+1)^\circ} &= \frac{1}{\sin 1^\circ} \sum_{k=1}^{89} \left[\cot k^\circ - \cot(k+1)^\circ\right]\\ &= \frac{1}{\sin 1^\circ} \cdot \cot 1^\circ = \frac{\cos 1^\circ}{\sin^2 1^\circ},\end{aligned}$$

thus completing the proof.

2. [China 2001, by Xiaoyang Su] Let ABC be a triangle, and let x be a nonnegative real number. Prove that

$$a^x \cos A + b^x \cos B + c^x \cos C \le \frac{1}{2}(a^x + b^x + c^x).$$

Solution: By symmetry, we may assume that $a \ge b \ge c$. Hence $A \ge B \ge C$, and so $\cos A \le \cos B \le \cos C$. Thus

$$(a^x - b^x)(\cos A - \cos B) \le 0,$$

or

$$a^x \cos A + b^x \cos B \le a^x \cos B + b^x \cos A.$$

Adding the last inequality with its analogous cyclic symmetric forms and then adding $a^x \cos A + b^x \cos B + c^x \cos C$ to both sides of the resulting inequality gives

$$\begin{aligned}&3(a^x \cos A + b^x \cos B + c^x \cos C)\\ &\quad \le (a^x + b^x + c^x)(\cos A + \cos B + \cos C),\end{aligned}$$

from which the desired result follows as a consequence of Introductory Problem 27(b).

Note: The above solution is similar to the proof of **Chebyshev's inequality**. We can also apply the **rearrangement inequality** to simplify our work. Because $a \geq b \geq c$ and $\cos A \leq \cos B \leq \cos C$, we have

$$a^x \cos A + b^x \cos B + c^x \cos C \leq a^x \cos B + b^x \cos C + c^x \cos A$$

and

$$a^x \cos A + b^x \cos B + c^x \cos C \leq a^x \cos C + b^x \cos A + c^x \cos B.$$

Hence

$$\begin{aligned} &3(a^x \cos A + b^x \cos B + c^x \cos C) \\ &\leq (a^x + b^x + c^x)(\cos A + \cos B + \cos C). \end{aligned}$$

3. Let x, y, z be positive real numbers.

   (a) Prove that

   $$\frac{x}{\sqrt{1+x^2}} + \frac{y}{\sqrt{1+y^2}} + \frac{z}{\sqrt{1+z^2}} \leq \frac{3\sqrt{3}}{2}$$

   if $x + y + z = xyz$;

   (b) Prove that

   $$\frac{x}{1-x^2} + \frac{y}{1-y^2} + \frac{z}{1-z^2} \geq \frac{3\sqrt{3}}{2}$$

   if $0 < x, y, z < 1$ and $xy + yz + zx = 1$.

Solution: Both problems can be solved by trigonometric substitutions.

(a) By Introductory Problem 20(a), there is an acute triangle ABC with $\tan A = x$, $\tan B = y$, and $\tan C = z$. Note that

$$\frac{\tan A}{\sqrt{1+\tan^2 A}} = \frac{\tan A}{\sec A} = \sin A.$$

The desired inequality becomes

$$\sin A + \sin B + \sin C \leq \frac{3\sqrt{3}}{2},$$

which is Introductory Problem 28(c).

(b) From the given condition and Introductory Problem 19(a), we can assume that there is an acute triangle ABC such that

$$\tan\frac{A}{2} = x, \quad \tan\frac{B}{2} = y, \quad \tan\frac{C}{2} = z.$$

By the **double-angle formulas**, it suffices to prove that

$$\tan A + \tan B + \tan C \geq 3\sqrt{3},$$

which is Introductory Problem 20(b).

4. [China 1997] Let x, y, z be real numbers with $x \geq y \geq z \geq \frac{\pi}{12}$ such that $x + y + z = \frac{\pi}{2}$. Find the maximum and the minimum values of the product $\cos x \sin y \cos z$.

Solution: Let $p = \cos x \sin y \cos z$. Because $\frac{\pi}{2} \geq y \geq z$, $\sin(y - z) \geq 0$. By the **product-to-sum formulas**, we have

$$p = \frac{1}{2}\cos x[\sin(y + z) + \sin(y - z)] \geq \frac{1}{2}\cos x \sin(y + z) = \frac{1}{2}\cos^2 x.$$

Note that $x = \frac{\pi}{2} - (y + z) \leq \frac{\pi}{2} - 2 \cdot \frac{\pi}{12} = \frac{\pi}{3}$. Hence the minimum value of p is $\frac{1}{2}\cos^2\frac{\pi}{3} = \frac{1}{8}$, obtained when $x = \frac{\pi}{3}$ and $y = z = \frac{\pi}{12}$.
On the other hand, we also have

$$p = \frac{1}{2}\cos z[\sin(x + y) - \sin(x - y)] \leq \frac{1}{2}\cos^2 z,$$

by noting that $\sin(x - y) \geq 0$ and $\sin(x + y) = \cos z$. By the **double-angle formulas**, we deduce that

$$p \leq \frac{1}{4}(1 + \cos 2z) \leq \frac{1}{4}\left(1 + \cos\frac{\pi}{6}\right) = \frac{2 + \sqrt{3}}{8}.$$

This maximum value is obtained if and only if $x = y = \frac{5\pi}{24}$ and $z = \frac{\pi}{12}$.

5. Let ABC be an acute-angled triangle, and for $n = 1, 2, 3$, let

$$x_n = 2^{n-3}(\cos^n A + \cos^n B + \cos^n C) + \cos A \cos B \cos C.$$

Prove that

$$x_1 + x_2 + x_3 \geq \frac{3}{2}.$$

Solution: By the **arithmetic–geometric means inequality,**

$$\cos^3 x + \frac{\cos x}{4} \geq \cos^2 x$$

for x such that $\cos x \geq 0$. Because triangle ABC is acute, $\cos A$, $\cos B$, and $\cos C$ are nonnegative. Setting $x = A, x = B, x = C$ and adding the resulting inequalities yields

$$x_1 + x_3 \geq \cos^2 A + \cos^2 B + \cos^2 C + 2\cos A \cos B \cos C = 2x_2.$$

Consequently,

$$x_1 + x_2 + x_3 \geq 3x_2 = \frac{3}{2},$$

by Introductory Problem 24(d).

6. Find the sum of all x in the interval $[0, 2\pi]$ such that

$$3\cot^2 x + 8\cot x + 3 = 0.$$

Solution: Consider the quadratic equation

$$3u^2 + 8u + 3 = 0.$$

The roots of the above equation are $u_1 = \frac{-8+2\sqrt{7}}{6}$ and $u_2 = \frac{-8-2\sqrt{7}}{6}$. Both roots are real, and their product u_1u_2 is equal to -1 (by **Viète's theorem**).

Because $y = \cot x$ is a bijection from the interval $(0, \pi)$ to the real numbers, there is a unique pair of numbers $x_{1,1}$ and $x_{2,1}$ with $0 < x_{1,1}, x_{2,1} < \pi$ such that $\cot x_{1,1} = u_1$ and $\cot x_{2,1} = u_2$. Because u_1, u_2 are negative, $\frac{\pi}{2} < x_{1,1}, x_{2,1} < \pi$, and so $\pi < x_{1,1} + x_{2,1} < 2\pi$. Because $\cot x \tan x = 1$ and both $\tan x$ and $\cot x$ have period π, it follows that

$$1 = \cot x \tan x = \cot x \cot\left(\frac{\pi}{2} - x\right) = \cot x \cot\left(\frac{3\pi}{2} - x\right)$$
$$= \cot x_{1,1} \cot x_{2,1}.$$

Therefore, $x_{1,1}+x_{2,1} = \frac{3\pi}{2}$. Likewise, in the interval $(\pi, 2\pi)$, there is a unique pair of numbers $x_{1,2}$ and $x_{2,2}$ satisfying the conditions of the problem with $x_{1,2}+x_{2,2} = \frac{7\pi}{2}$. Thus the answer to the problem is $x_{1,1}+x_{2,1}+x_{1,2}+x_{2,2} = 5\pi$.

7. Let ABC be an acute-angled triangle with area K. Prove that

$$\sqrt{a^2b^2-4K^2}+\sqrt{b^2c^2-4K^2}+\sqrt{c^2a^2-4K^2}=\frac{a^2+b^2+c^2}{2}.$$

Solution: We have $2K = ab\sin C = bc\sin A = ca\sin B$. The expression on the left-hand side of the desired equation is equal to

$$\begin{aligned}
&\sqrt{a^2b^2-a^2b^2\sin^2 C}+\sqrt{b^2c^2-b^2c^2\sin^2 A}+\sqrt{c^2a^2-c^2a^2\sin^2 B}\\
&= ab\cos C+bc\cos A+ca\cos B\\
&= \frac{a}{2}(b\cos C+c\cos B)+\frac{b}{2}(c\cos A+a\cos C)\\
&\quad+\frac{c}{2}(a\cos B+b\cos A)\\
&= \frac{a}{2}\cdot a+\frac{b}{2}\cdot b+\frac{c}{2}\cdot c,
\end{aligned}$$

and the conclusion follows.

Note: We encourage the reader to explain why this problem is the equality case of Advanced Problem 42(a).

8. Compute the sums

$$\binom{n}{1}\sin a+\binom{n}{2}\sin 2a+\cdots+\binom{n}{n}\sin na$$

and

$$\binom{n}{1}\cos a+\binom{n}{2}\cos 2a+\cdots+\binom{n}{n}\cos na.$$

Solution: Let S_n and T_n denote the first and second sums, respectively. Set the complex number $z=\cos a+i\sin a$. Then, by **de Moivre's formula**, we have $z^n=\cos na+i\sin na$. By the **binomial theorem**, we obtain

$$\begin{aligned}
1+T_n+iS_n &= 1+\binom{n}{1}(\cos a+i\sin a)+\binom{n}{2}(\cos 2a+i\sin 2a)\\
&\quad+\cdots+\binom{n}{n}(\cos na+i\sin na)\\
&= \binom{n}{0}z^0+\binom{n}{1}z+\binom{n}{2}z^2+\cdots+\binom{n}{n}z^n\\
&= (1+z)^n.
\end{aligned}$$

Because

$$1 + z = 1 + \cos a + i \sin a = 2\cos^2 \frac{a}{2} + 2i \sin \frac{a}{2} \cos \frac{a}{2}$$
$$= 2\cos \frac{a}{2} \left(\cos \frac{a}{2} + i \sin \frac{a}{2} \right),$$

it follows that

$$(1+z)^n = 2^n \cos^n \frac{a}{2} \left(\cos \frac{na}{2} + i \sin \frac{na}{2} \right),$$

again by de Moivre's formula. Therefore,

$$(1 + T_n) + i S_n = \left(2^n \cos^n \frac{a}{2} \cos \frac{na}{2} \right) + i \left(2^n \cos^n \frac{a}{2} \sin \frac{na}{2} \right),$$

and so

$$S_n = 2^n \cos^n \frac{a}{2} \sin \frac{na}{2} \quad \text{and} \quad T_n = -1 + 2^n \cos^n \frac{a}{2} \cos \frac{na}{2}.$$

9. [Putnam 2003] Find the minimum value of

$$|\sin x + \cos x + \tan x + \cot x + \sec x + \csc x|$$

for real numbers x.

Solution: Set $a = \sin x$ and $b = \cos x$. We want to minimize

$$P = \left| a + b + \frac{a}{b} + \frac{b}{a} + \frac{1}{a} + \frac{1}{b} \right|$$
$$= \left| \frac{ab(a+b) + a^2 + b^2 + a + b}{ab} \right|.$$

Note that $a^2 + b^2 = \sin^2 x + \cos^2 x = 1$. Set $c = a + b$. Then $c^2 = (a+b)^2 = 1 + 2ab$, and so $2ab = c^2 - 1$. Note also that by the **addition and subtraction formulas**, we have

$$c = \sin x + \cos x = \sqrt{2}\left(\frac{\sqrt{2}}{2} \sin x + \frac{\sqrt{2}}{2} \cos x \right) = \sqrt{2} \sin\left(\frac{\pi}{4} + x \right),$$

and so the range of c is the interval $[-\sqrt{2}, \sqrt{2}]$. Consequently, it suffices to find the minimum of

$$\begin{aligned} P(c) &= \left|\frac{2ab(a+b)+2+2(a+b)}{2ab}\right| \\ &= \left|\frac{c(c^2-1)+2(c+1)}{c^2-1}\right| = \left|c+\frac{2}{c-1}\right| \\ &= \left|c-1+\frac{2}{c-1}+1\right|. \end{aligned}$$

for c in the interval $[-\sqrt{2}, \sqrt{2}]$. If $c-1>0$, then by the **arithmetic–geometric means inequality**, $(c-1)+\frac{2}{c-1} > 2\sqrt{2}$, and so $P(c) > 1+2\sqrt{2}$. If $c-1<0$, then by the same token,

$$(c-1)+\frac{2}{c-1} = -\left((1-c)+\frac{2}{1-c}\right) \leq -2\sqrt{2},$$

with equality if and only if $1-c=\frac{2}{1-c}$, or $c=1-\sqrt{2}$. It follows that the minimum value sought is $\left|-2\sqrt{2}+1\right| = 2\sqrt{2}-1$, obtained when $c=1-\sqrt{2}$.

Note: Taking the derivative of the function

$$f(x) = \sin x + \cos x + \tan x + \cot x + \sec x + \csc x$$

and considering only its critical points is a troublesome approach to this problem, because it is difficult to show that $f(x)$ does not cross the x axis smoothly. Indeed, with a little bit more work, we can show that $f(x) \neq 0$ with the presented solution.

10. [Belarus 1999] Two real sequences $x_1, x_2, \ldots$ and $y_1, y_2, \ldots$ are defined in the following way:

$$x_1 = y_1 = \sqrt{3}, \quad x_{n+1} = x_n + \sqrt{1+x_n^2}, \quad y_{n+1} = \frac{y_n}{1+\sqrt{1+y_n^2}},$$

for all $n \geq 1$. Prove that $2 < x_n y_n < 3$ for all $n > 1$.

Solution: Writing $x_n = \tan a_n$ for $0^\circ < a_n < 90^\circ$, by the **half-angle formula** we have

$$\begin{aligned} x_{n+1} &= \tan a_n + \sqrt{1+\tan^2 a_n} = \tan a_n + \sec a_n \\ &= \frac{1+\sin a_n}{\cos a_n} = \tan\left(\frac{90^\circ + a_n}{2}\right). \end{aligned}$$

Because $a_1 = 60°$, we have $a_2 = 75°$, $a_3 = 82.5°$, and in general $a_n = 90° - \frac{30°}{2^{n-1}}$. Thus

$$x_n = \tan\left(90° - \frac{30°}{2^{n-1}}\right) = \cot\left(\frac{30°}{2^{n-1}}\right) = \cot\theta_n,$$

where $\theta_n = \frac{30°}{2^{n-1}}$.

A similar calculation shows that

$$y_n = \tan 2\theta_n = \frac{2\tan\theta_n}{1-\tan^2\theta_n},$$

implying that

$$x_n y_n = \frac{2}{1-\tan^2\theta_n}.$$

Because $0° < \theta_n < 45°$, we have $0 < \tan^2\theta_n < 1$ and $x_n y_n > 2$. For $n > 1$, we have $\theta_n < 30°$, implying that $\tan^2\theta_n < \frac{1}{3}$ and $x_n y_n < 3$.

11.. Let a, b, c be real numbers such that

$$\sin a + \sin b + \sin c \geq \frac{3}{2}.$$

Prove that

$$\sin\left(a - \frac{\pi}{6}\right) + \sin\left(b - \frac{\pi}{6}\right) + \sin\left(c - \frac{\pi}{6}\right) \geq 0.$$

Solution: Assume for contradiction that

$$\sin\left(a - \frac{\pi}{6}\right) + \sin\left(b - \frac{\pi}{6}\right) + \sin\left(c - \frac{\pi}{6}\right) < 0.$$

Then by the **addition and subtraction formulas**, we have

$$\frac{1}{2}(\cos a + \cos b + \cos c) > \frac{\sqrt{3}}{2}(\sin a + \sin b + \sin c) \geq \frac{3\sqrt{3}}{4}.$$

It follows that

$$\cos a + \cos b + \cos c > \frac{3\sqrt{3}}{2},$$

which implies that

$$\sin\left(a+\frac{\pi}{3}\right)+\sin\left(b+\frac{\pi}{3}\right)+\sin\left(c+\frac{\pi}{3}\right)$$
$$=\frac{1}{2}(\sin a+\sin b+\sin c)+\frac{\sqrt{3}}{2}(\cos a+\cos b+\cos c)$$
$$>\frac{1}{2}\cdot\frac{3}{2}+\frac{\sqrt{3}}{2}\cdot\frac{3\sqrt{3}}{2}=3,$$

which is impossible, because $\sin x < 1$.

12. Consider any four numbers in the interval $\left[\frac{\sqrt{2}-\sqrt{6}}{2}, \frac{\sqrt{2}+\sqrt{6}}{2}\right]$. Prove that there are two of them, say a and b, such that

$$\left|a\sqrt{4-b^2}-b\sqrt{4-a^2}\right| \le 2.$$

Solution: Dividing both sides of the inequality by 4 yields

$$\left|\frac{a}{2}\sqrt{1-\left(\frac{b}{2}\right)^2}-\frac{b}{2}\sqrt{1-\left(\frac{a}{2}\right)^2}\right| \le \frac{1}{2}.$$

We substitute $\frac{a}{2}=\sin x$ and $\frac{b}{2}=\sin y$. The last inequality reduces to

$$|\sin(x-y)| = |\sin x\cos y-\sin y\cos x| \le \sin\frac{\pi}{6}. \qquad (*)$$

We want to find t_1 and t_2 such that

$$\sin t_1=\frac{\sqrt{2}-\sqrt{6}}{4} \quad \text{and} \quad \sin t_2=\frac{\sqrt{2}+\sqrt{6}}{4}.$$

By the **double-angle formulas**, we conclude that $\cos 2t_1 = 1-2\sin^2 t_1 = 1-\frac{8-4\sqrt{3}}{8}=\frac{\sqrt{3}}{2}=\cos\left(\pm\frac{\pi}{6}\right)$ and $\cos 2t_2=-\frac{\sqrt{3}}{2}=\cos\frac{5\pi}{6}$. Because $y=\sin x$ is a **one-to-one** and **onto** map between the intervals $\left[-\frac{\pi}{2},\frac{\pi}{2}\right]$ and $[-1,1]$, it follows that $t_1=-\frac{\pi}{12}$ and $t_2=\frac{5\pi}{12}$.

We divide the interval $\left[-\frac{\pi}{12},\frac{5\pi}{12}\right]$ into three disjoint intervals of length $\frac{\pi}{6}$: $I_1=\left[-\frac{\pi}{12},\frac{\pi}{12}\right)$, $I_2=\left[\frac{\pi}{12},\frac{\pi}{4}\right)$, and $I_3=\left[\frac{\pi}{4},\frac{5\pi}{12}\right]$. The function $y=2\sin x$ takes the intervals I_1, I_2, I_3 injectively and surjectively to the intervals $I_1' =$

$\left[\frac{\sqrt{2}-\sqrt{6}}{2}, 2\sin\frac{\pi}{12}\right)$, $I_2 = \left[2\sin\frac{\pi}{12}, \sqrt{2}\right)$, $I_3' = \left[\sqrt{2}, \frac{\sqrt{2}+\sqrt{6}}{2}\right]$, respectively. By the **pigeonhole principle**, one of the intervals I_1', I_2', or I_3' contains two of the four given numbers, say a and b. It follows that one of the intervals I_1, I_2, or I_3 contains x and y such that $a = 2\sin x$ and $b = 2\sin y$. Because the intervals I_1, I_2, and I_3 have equal lengths of $\frac{\pi}{6}$, it follows that $|x - y| \le \frac{\pi}{6}$. We have obtained the desired the inequality $(*)$.

13. Let a and b be real numbers in the interval $[0, \frac{\pi}{2}]$. Prove that

$$\sin^6 a + 3\sin^2 a\cos^2 b + \cos^6 b = 1$$

if and only if $a = b$.

Solution: The first equality can be rewritten as

$$(\sin^2 a)^3 + (\cos^2 b)^3 + (-1)^3 - 3(\sin^2 a)(\cos^2 b)(-1) = 0. \qquad (*)$$

We will use the identity

$$x^3 + y^3 + z^3 - 3xyz = \frac{1}{2}(x + y + z)[(x - y)^2 + (y - z)^2 + (z - x)^2].$$

Let $x = \sin^2 a$, $y = \cos^2 b$, and $z = -1$. According to equation $(*)$ we have $x^3+y^3+z^3-3xyz = 0$. Hence $x+y+z = 0$ or $(x-y)^2+(y-z)^2+(z-x)^2 = 0$. The latter would imply $x = y = z$, or $\sin^2 a = \cos^2 b = -1$, which is impossible. Thus $x + y + z = 0$, so that $\sin^2 a + \cos^2 b - 1 = 0$, or $\sin^2 a = 1 - \cos^2 b$. It follows that $\sin^2 a = \sin^2 b$, and taking into account that $0 \le a, b \le \frac{\pi}{2}$, we obtain $a = b$.

Even though all the steps above are reversible, we will show explicitly that if $a = b$, then

$$\sin^6 a + \cos^6 a + 3\sin^2 a\cos^2 b = 1.$$

Indeed, the expression on the left-hand side could be written as

$$(\sin^2 a + \cos^2 a)(\sin^4 a - \sin^2 a\cos^2 a + \cos^4 a) + 3\sin^2 a\cos^2 a$$
$$= (\sin^2 a + \cos^2 a)^2 - 3\sin^2 a\cos^2 a + 3\sin^2 a\cos^2 a = 1.$$

14. Let x, y, z be real numbers with $0 < x < y < z < \frac{\pi}{2}$. Prove that

$$\frac{\pi}{2} + 2\sin x\cos y + 2\sin y\cos z \ge \sin 2x + \sin 2y + \sin 2z.$$

Solution: By the **double-angle formulas**, the above inequalities reduce to

$$\frac{\pi}{2} > 2\sin x(\cos x - \cos y) + 2\sin y(\cos y - \cos z) + 2\sin z\cos z,$$

or

$$\frac{\pi}{4} > \sin x(\cos x - \cos y) + \sin y(\cos y - \cos z) + \sin z\cos z.$$

As shown in Figure 5.1, in the rectangular coordinate plane, we consider points $O = (0, 0)$, $A = (\cos x, \sin x)$, $A_1 = (\cos x, 0)$, $B = (\cos y, \sin y)$, $B_1 = (\cos y, 0)$, $B_2 = (\cos y, \sin x)$, $C = (\cos z, \sin z)$, $C_1 = (\cos z, 0)$, $C_2 = (\cos z, \sin y)$, and $D = (0, \sin z)$. Points A, B, and C are in the first quadrant of the coordinate plane, and they lie on the unit circle in counterclockwise order.

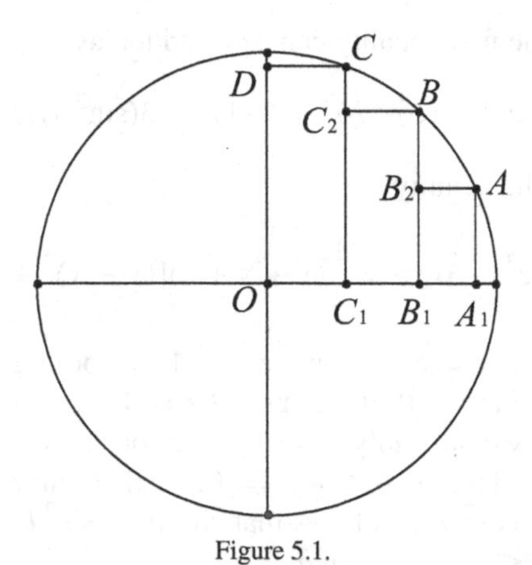

Figure 5.1.

Let $\mathcal{D}$ denote the region enclosed by the unit circle in the first quadrant (including the boundary). It is not difficult to see that quadrilaterals $AA_1B_1B_2$, $BB_1C_1C_2$, and CC_1OD are nonoverlapping rectangles inside region $\mathcal{D}$. It is also not difficult to see that $[\mathcal{D}] = \frac{\pi}{4}$, $[AA_1B_1B_2] = \sin x(\cos x - \cos y)$, $[BB_1C_1C_2] = \sin y(\cos y - \cos z)$, and $[CC_1OD] = \sin z\cos z$, from which our desired result follows.

15. For a triangle XYZ, let r_{XYZ} denote its inradius. Given that the convex pentagon $ABCDE$ is inscribed in a circle, prove that if $r_{ABC} = r_{AED}$ and $r_{ABD} = r_{AEC}$, then triangles ABC and AED are congruent.

Solution: Let R be the radius of the circle in which $ABCDE$ is inscribed.

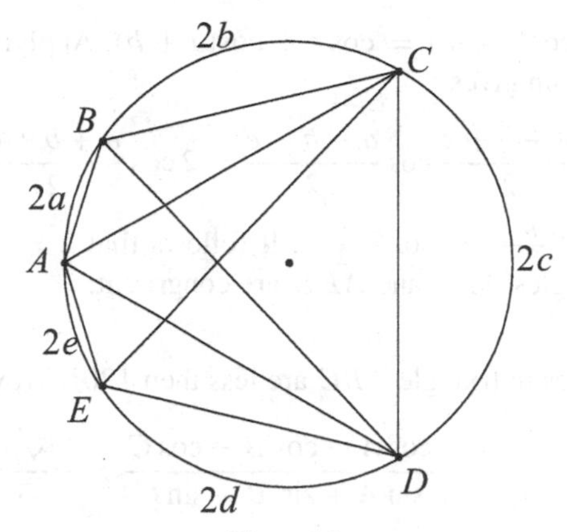

Figure 5.2.

As shown in the proof of Problem 27(a), if ABC is a triangle with inradius r and circumradius R, then

$$1 + \frac{r}{R} = \cos A + \cos B + \cos C = \cos A - \cos(A + C) + \cos C.$$

Let $2a$, $2b$, $2c$, $2d$, and $2e$ be the measures of arcs $\widehat{AB}$, $\widehat{BC}$, $\widehat{CD}$, $\widehat{DE}$, and $\widehat{EA}$, respectively. Then $a + b + c + d + e = 180^\circ$. Because $r_{ABC} = r_{AED}$ and $r_{ABD} = r_{AEC}$, we have

$$\cos a - \cos(a + b) + \cos b = \cos d + \cos e - \cos(d + e) \qquad (*)$$

and

$$\cos a + \cos(b + c) + \cos(d + e) = \cos e + \cos(c + d) + \cos(a + b).$$

Subtracting the two equations, we obtain $\cos b + \cos(c+d) = \cos d + \cos(b + c)$, or

$$2\cos\frac{b+c+d}{2}\cos\frac{b-c-d}{2} = 2\cos\frac{b+c+d}{2}\cos\frac{d-b-c}{2}$$

by **sum-to-product formulas**. It follows that

$$\cos\frac{b-c-d}{2} = \cos\frac{d-b-c}{2},$$

and so $b = d$. Plugging this result into equation $(*)$ yields

$$\cos a - \cos(a + b) + \cos b = \cos b + \cos e - \cos(b + e),$$

or $\cos a + \cos(b+e) = \cos e + \cos(a+b)$. Applying the sum-to-product formulas again gives

$$2\cos\frac{a+b+e}{2}\cos\frac{a-b-e}{2} = 2\cos\frac{a+b+e}{2}\cos\frac{e-a-b}{2},$$

and so $\cos\frac{a-b-e}{2} = \cos\frac{e-a-b}{2}$. It follows that $a = e$. Because $a = e$ and $b = d$, triangles ABC and AED are congruent.

16. All the angles in triangle ABC are less then $120°$. Prove that

$$\frac{\cos A + \cos B - \cos C}{\sin A + \sin B - \sin C} > -\frac{\sqrt{3}}{3}.$$

Solution: Consider the triangle $A_1B_1C_1$, as shown in Figure 5.3, where $\angle A_1 = 120° - \angle A$, $\angle B_1 = 120° - \angle B$, and $\angle C_1 = 120° - \angle C$. The given condition guarantees the existence of such a triangle.

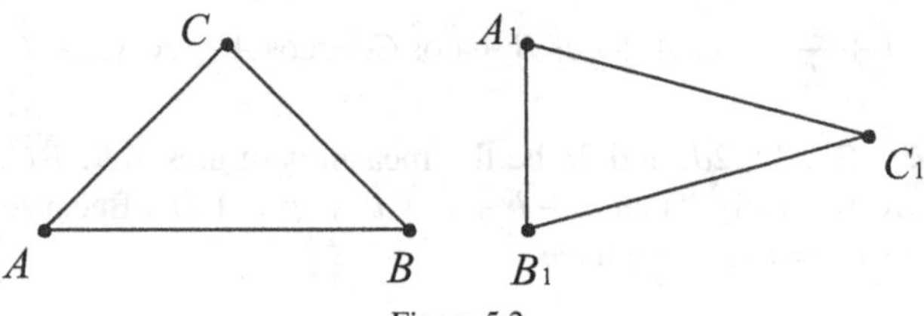

Figure 5.3.

Applying the **triangle inequality** in triangle $A_1B_1C_1$ gives $B_1C_1 + C_1A_1 > A_1B_1$; that is

$$\sin A_1 + \sin B_1 > \sin C_1$$

by applying the **law of sines** to triangle $A_1B_1C_1$. It follows that

$$\sin(120° - A) + \sin(120° - B) > \sin(120° - C),$$

or

$$\frac{\sqrt{3}}{2}(\cos A + \cos B - \cos C) + \frac{1}{2}(\sin A + \sin B - \sin C) > 0.$$

Taking into account that $a + b > c$ implies $\sin A + \sin B - \sin C > 0$, the above inequality can be rewritten as

$$\frac{\sqrt{3}}{2}\cdot\frac{\cos A + \cos B - \cos C}{\sin A + \sin B - \sin C} + \frac{1}{2} > 0,$$

from which the conclusion follows.

17. [USAMO 2002] Let ABC be a triangle such that

$$\left(\cot\frac{A}{2}\right)^2+\left(2\cot\frac{B}{2}\right)^2+\left(3\cot\frac{C}{2}\right)^2=\left(\frac{6s}{7r}\right)^2,$$

where s and r denote its semiperimeter and its inradius, respectively. Prove that triangle ABC is similar to a triangle T whose side lengths are all positive integers with no common divisor and determine these integers.

Solution: Define

$$u=\cot\frac{A}{2},\quad v=\cot\frac{B}{2},\quad w=\cot\frac{C}{2}.$$

As shown in Figure 5.4, denote the incenter of triangle ABC by I, and let D, E, and F be the points of tangency of the incircle with sides BC, CA, and AB, respectively. Then $|EI|=r$, and by the standard formula, $|AE|=s-a$.

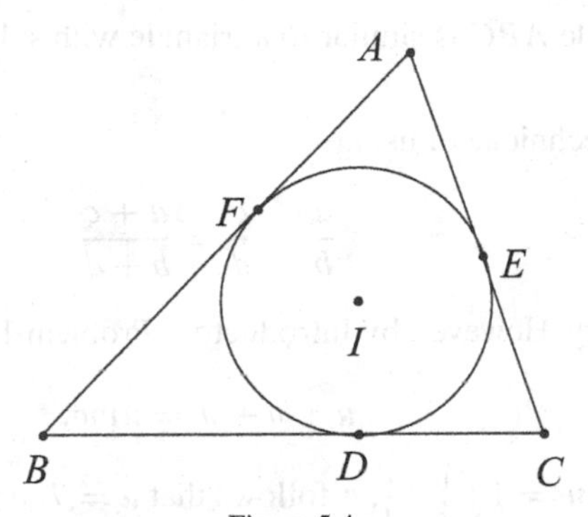

Figure 5.4.

We have

$$u=\cot\frac{A}{2}=\frac{|AE|}{|EI|}=\frac{s-a}{r},$$

and similarly $v=\frac{s-b}{r}$, $w=\frac{s-c}{r}$. Because

$$\frac{s}{r}=\frac{(s-a)+(s-b)+(s-c)}{r}=u+v+w,$$

we can rewrite the given relation as

$$49\left[u^2+4v^2+9w^2\right]=36(u+v+w)^2.$$

Expanding the last equality and canceling like terms, we obtain

$$13u^2 + 160v^2 + 405w^2 - 72(uv + vw + wu) = 0,$$

or

$$(3u - 12v)^2 + (4v - 9w)^2 + (18w - 2u)^2 = 0.$$

Therefore, $u : v : w = 1 : \frac{1}{4} : \frac{1}{9}$. This can also be realized by recognizing that the given relation corresponds to equality in **Cauchy–Schwarz inequality**

$$(6^2 + 3^2 + 2^2)\left[u^2 + (2v)^2 + (3w)^2\right] \geq (6 \cdot u + 3 \cdot 2v + 2 \cdot 3w)^2.$$

After multiplying by r, we see that

$$\frac{s-a}{36} = \frac{s-b}{9} = \frac{s-c}{4} = \frac{2s-b-c}{9+4} = \frac{2s-c-a}{4+36} = \frac{2s-a-b}{36+9}$$
$$= \frac{a}{13} = \frac{b}{40} = \frac{c}{45};$$

that is, triangle ABC is similar to a triangle with side lengths 13, 40, 45.

Note: The technique of using

$$\frac{a}{b} = \frac{c}{d} = \frac{a+c}{b+d}$$

is rather tricky. However, by Introductory Problem 19(a), we can have

$$u + v + w = uvw.$$

Since $u : v : w = 1 : \frac{1}{4} : \frac{1}{9}$, it follows that $u = 7$, $v = \frac{7}{4}$, and $w = \frac{7}{9}$. Hence by the **double-angle formulas**, $\sin A = \frac{7}{25}$, $\sin B = \frac{56}{65}$, and $\sin C = \frac{63}{65}$, or

$$\sin A = \frac{13}{\frac{325}{7}}, \quad \sin B = \frac{40}{\frac{325}{7}}, \quad \sin C = \frac{45}{\frac{325}{7}}.$$

By the **extended law of sines**, triangle ABC is similar to triangle T with side lengths 13, 40, and 45. (The circumcircle of T has diameter $\frac{325}{7}$.)

18. [USAMO 1996] Prove that the average of the numbers

$$2\sin 2^\circ, \quad 4\sin 4^\circ, \quad 6\sin 6^\circ, \quad \ldots, \quad 180\sin 180^\circ,$$

is $\cot 1^\circ$.

First Solution: We need to prove that

$$2\sin 2^\circ + 4\sin 4^\circ + \cdots + 178\sin 178^\circ = 90\cot 1^\circ,$$

which is equivalent to

$$\begin{aligned}&2\sin 2^\circ \cdot \sin 1^\circ + 2(2\sin 4^\circ \cdot \sin 1^\circ) + \cdots + 89(2\sin 178^\circ \cdot \sin 1^\circ)\\ &\quad = 90\cos 1^\circ.\end{aligned}$$

Note that

$$2\sin 2k^\circ \sin 1^\circ = \cos(2k-1)^\circ - \cos(2k+1)^\circ.$$

We have

$$\begin{aligned}&2\sin 2^\circ \cdot \sin 1^\circ + 2(2\sin 4^\circ \cdot \sin 1^\circ) + \cdots + 89(2\sin 178^\circ \cdot \sin 1^\circ)\\ &\quad = (\cos 1^\circ - \cos 3^\circ) + 2(\cos 3^\circ - \cos 5^\circ)\\ &\qquad + \cdots + 89(\cos 177^\circ - \cos 179^\circ)\\ &\quad = \cos 1^\circ + \cos 3^\circ + \cdots + \cos 177^\circ - 89\cos 179^\circ\\ &\quad = \cos 1^\circ + (\cos 3^\circ + \cos 177^\circ) + \cdots + (\cos 89^\circ + \cos 91^\circ)\\ &\qquad + 89\cos 1^\circ\\ &\quad = \cos 1^\circ + 89\cos 1^\circ = 90\cos 1^\circ,\end{aligned}$$

as desired.

Note: The techniques of telescoping sum and pairing of summands involved in the first solution is rather tricky. The second solution involves complex numbers. It is slightly longer than the first solution. But for the reader who is familiar with **the rules of operation for complex numbers** and **geometric series**, every step is natural.

Second Solution: Set the complex number $z = \cos 2^\circ + i\sin 2^\circ$. Then, by **de Moivre's formula**, we have $z^n = \cos 2n^\circ + i\sin 2n^\circ$. Let a and b be real numbers such that

$$z + 2z^2 + \cdots + 89z^{89} = a + bi.$$

Because $\sin 180^\circ = 0$,

$$b = \frac{1}{2}(2\sin 2^\circ + 4\sin 4^\circ + \cdots + 178\sin 178^\circ + 180\sin 180^\circ),$$

and it suffices to show that $b = 45\cot 1^\circ$.

Set

$$p_n(x) = x + 2x^2 + \cdots + nx^n.$$

Then

$$(1-x)p_n(x) = p_n(x) - xp_n(x) = x + x^2 + \cdots + x^n - nx^{n+1}.$$

Set

$$q_n(x) = (1-x)p_n(x) + nx^{n+1} = x + x^2 + \cdots + x^n.$$

Then $(1-x)q_n(x) = q_n(x) - xq_n(x) = x - x^{n+1}$. Consequently, we have

$$p_n(x) = \frac{q_n(x)}{1-x} - \frac{nx^{n+1}}{1-x} = \frac{x - x^{n+1}}{(1-x)^2} - \frac{nx^{n+1}}{1-x}.$$

It follows that

$$\begin{aligned} a + bi &= z + 2z^2 + \cdots + 89z^{89} = p_{89}(z) \\ &= \frac{z - z^{90}}{(1-z)^2} - \frac{89z^{90}}{1-z} = \frac{z+1}{(z-1)^2} - \frac{89}{z-1}, \end{aligned}$$

because $z^{90} = \cos 180° + i \sin 180° = -1$. Note that $z + 1 = \operatorname{cis} 2° + \operatorname{cis} 0° = 2\cos 1° \operatorname{cis} 1°$ and $z - 1 = \operatorname{cis} 2° - \operatorname{cis} 0° = 2 \sin 1° \operatorname{cis} 91°$, and so

$$\begin{aligned} a + bi &= \frac{2\cos 1° \operatorname{cis} 1°}{(2\sin 1° \operatorname{cis} 91°)^2} - \frac{89}{2\sin 1° \operatorname{cis} 91°} \\ &= \frac{2\cos 1° \operatorname{cis} 1°}{4\sin^2 1° \operatorname{cis} 182°} - \frac{89 \operatorname{cis}(-91°)}{2\sin 1°} \\ &= \frac{\cos 1° \operatorname{cis}(-181°)}{2\sin^2 1°} - \frac{89 \operatorname{cis}(-91°)}{2\sin 1°}. \end{aligned}$$

Therefore,

$$\begin{aligned} b &= \frac{\cos 1° \sin(-181°)}{2\sin^2 1°} - \frac{89 \sin(-91°)}{2\sin 1°} = \frac{\cos 1° \sin 1°}{2\sin^2 1°} + \frac{89\cos 1°}{2\sin 1°} \\ &= \frac{\cos 1°}{2\sin 1°} + \frac{89\cos 1°}{2\sin 1°} = 45 \cot 1°, \end{aligned}$$

as desired.

19. Prove that in any acute triangle ABC,

$$\cot^3 A + \cot^3 B + \cot^3 C + 6\cot A \cot B \cot C \geq \cot A + \cot B + \cot C.$$

Solution: Let $\cot A = x$, $\cot B = y$, and $\cot C = z$. Because $xy+yz+zx = 1$ (Introductory Problem 21), it suffices to prove the homogeneous inequality

$$x^3 + y^3 + z^3 + 6xyz \geq (x + y + z)(xy + yz + zx).$$

But this is equivalent to

$$x(x - y)(x - z) + y(y - z)(y - x) + z(z - x)(z - y) \geq 0,$$

which is **Schur's inequality**.

20. [Turkey 1998] Let $\{a_n\}$ be the sequence of real numbers defined by $a_1 = t$ and $a_{n+1} = 4a_n(1 - a_n)$ for $n \geq 1$. For how many distinct values of t do we have $a_{1998} = 0$?

Solution: Let $f(x) = 4x(1-x) = 1-(2x-1)^2$. Observe that if $0 \leq f(x) \leq 1$, then $0 \leq x \leq 1$. Hence if $a_{1998} = 0$, then we must have $0 \leq t \leq 1$. Now choose $0 \leq \theta \leq \frac{\pi}{2}$ such that $\sin\theta = \sqrt{t}$. Observe that for any $\phi \in \mathbb{R}$,

$$f(\sin^2\phi) = 4\sin^2\phi\,(1 - \sin^2\phi) = 4\sin^2\phi\cos^2\phi = \sin^2 2\phi;$$

since $a_1 = \sin^2\theta$, it follows that

$$a_2 = \sin^2 2\theta, \quad a_3 = \sin^2 4\theta, \quad \ldots, \quad a_{1998} = \sin^2 2^{1997}\theta.$$

Therefore, $a_{1998} = 0$ if and only if $\sin 2^{1997}\theta = 0$. That is, $\theta = \frac{k\pi}{2^{1997}}$ for some integers k, and so the values of t for which $a_{1998} = 0$ are $\sin^2(k\pi/2^{1997})$, where $k \in \mathbb{Z}$. Therefore we get $2^{1996} + 1$ such values of t, namely, $\sin^2(k\pi/2^{1997})$ for $k = 0, 1, 2, \cdots, 2^{1996}$.

21. Triangle ABC has the following property: there is an interior point P such that $\angle PAB = 10°$, $\angle PBA = 20°$, $\angle PCA = 30°$, and $\angle PAC = 40°$. Prove that triangle ABC is isosceles.

Solution: Consider Figure 5.5, in which all angles are in degrees.

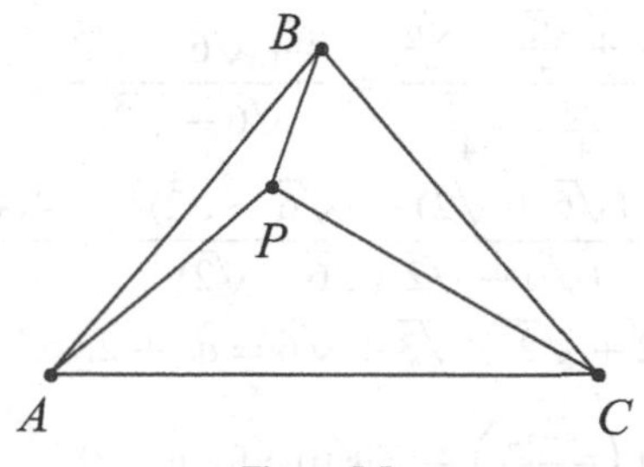

Figure 5.5.

Let $x = \angle PCB$ (in degrees). Then $\angle PBC = 80^\circ - x$. By the **law of sines** or by **Ceva's theorem**,

$$1 = \frac{PA}{PB} \cdot \frac{PB}{PC} \cdot \frac{PC}{PA} = \frac{\sin \angle PBA}{\sin \angle PAB} \cdot \frac{\sin \angle PCB}{\sin \angle PBC} \cdot \frac{\sin \angle PAC}{\sin \angle PCA}$$
$$= \frac{\sin 20^\circ \sin x \sin 40^\circ}{\sin 10^\circ \sin(80^\circ - x) \sin 30^\circ} = \frac{4 \sin x \sin 40^\circ \cos 10^\circ}{\sin(80^\circ - x)}.$$

The **product-to-sum formulas** yield

$$1 = \frac{2 \sin x (\sin 30^\circ + \sin 50^\circ)}{\sin(80^\circ - x)} = \frac{\sin x (1 + 2\cos 40^\circ)}{\sin(80^\circ - x)},$$

and so

$$2 \sin x \cos 40^\circ = \sin(80^\circ - x) - \sin x = 2\sin(40^\circ - x)\cos 40^\circ,$$

by the **difference-to-product formulas**. We conclude that $x = 40^\circ - x$, or $x = 20^\circ$. It follows that $\angle ACB = 50^\circ = \angle BAC$, and so triangle ABC is isosceles.

22. Let $a_0 = \sqrt{2} + \sqrt{3} + \sqrt{6}$, and let $a_{n+1} = \frac{a_n^2 - 5}{2(a_n + 2)}$ for integers $n > 0$. Prove that

$$a_n = \cot\left(\frac{2^{n-3}\pi}{3}\right) - 2$$

for all n.

Solution: By either the **double-angle** or the **half-angle formulas**, we obtain

$$\cot \frac{\pi}{24} = \frac{\cos \frac{\pi}{24}}{\sin \frac{\pi}{24}} = \frac{2\cos^2 \frac{\pi}{24}}{2 \sin \frac{\pi}{24} \cos \frac{\pi}{24}} = \frac{1 + \cos \frac{\pi}{12}}{\sin \frac{\pi}{12}}$$
$$= \frac{1 + \cos\left(\frac{\pi}{3} - \frac{\pi}{4}\right)}{\sin\left(\frac{\pi}{3} - \frac{\pi}{4}\right)} = \frac{1 + \cos \frac{\pi}{3} \cos \frac{\pi}{4} + \sin \frac{\pi}{3} \sin \frac{\pi}{4}}{\sin \frac{\pi}{3} \cos \frac{\pi}{4} - \cos \frac{\pi}{3} \sin \frac{\pi}{4}}$$
$$= \frac{1 + \frac{\sqrt{2}}{4} + \frac{\sqrt{6}}{4}}{\frac{\sqrt{6}}{4} - \frac{\sqrt{2}}{4}} = \frac{4 + \sqrt{6} + \sqrt{2}}{\sqrt{6} - \sqrt{2}}$$
$$= \frac{4(\sqrt{6} + \sqrt{2}) + (\sqrt{6} + \sqrt{2})^2}{(\sqrt{6} - \sqrt{2})(\sqrt{6} + \sqrt{2})} = \frac{4(\sqrt{6} + \sqrt{2}) + 8 + 4\sqrt{3}}{4}$$
$$= 2 + \sqrt{2} + \sqrt{3} + \sqrt{6} = a_0 + 2.$$

Hence $a_n = \cot\left(\frac{2^{n-3}\pi}{3}\right) - 2$ is true for $n = 0$.

It suffices to show that $b_n = \cot\left(\frac{2^{n-3}\pi}{3}\right)$, where $b_n = a_n + 2$, $n \geq 1$. The recursive relation becomes

$$b_{n+1} - 2 = \frac{(b_n - 2)^2 - 5}{2b_n},$$

or

$$b_{n+1} = \frac{b_n^2 - 1}{2b_n}.$$

Assuming, inductively, that $b_k = \cot c_k$, where $c_k = \frac{2^{k-3}\pi}{3}$, yields

$$b_{k+1} = \frac{\cot^2 c_k - 1}{2\cot c_k} = \cot 2c_k = \cot c_{k+1},$$

and we are done.

23. [APMC 1982] Let n be an integer with $n \geq 2$. Prove that

$$\prod_{k=1}^{n} \tan\left[\frac{\pi}{3}\left(1 + \frac{3^k}{3^n - 1}\right)\right] = \prod_{k=1}^{n} \cot\left[\frac{\pi}{3}\left(1 - \frac{3^k}{3^n - 1}\right)\right].$$

Solution: Let

$$u_k = \tan\left[\frac{\pi}{3}\left(1 + \frac{3^k}{3^n - 1}\right)\right] \text{ and } v_k = \tan\left[\frac{\pi}{3}\left(1 - \frac{3^k}{3^n - 1}\right)\right].$$

The desired equality becomes

$$\prod_{k=1}^{n} u_k v_k = 1. \tag{$*$}$$

Set

$$t_k = \tan\frac{3^{k-1}\pi}{3^n - 1}.$$

Applying the **addition and and subtraction formulas** yields

$$u_k = \tan\left(\frac{\pi}{3} + \frac{3^{k-1}\pi}{3^n - 1}\right) = \frac{\sqrt{3} + t_k}{1 - \sqrt{3}t_k} \quad \text{and} \quad v_k = \frac{\sqrt{3} - t_k}{1 + \sqrt{3}t_k}.$$

The **triple-angle formulas** give

$$t_{k+1} = \frac{3t_k - t_k^3}{1 - 3t_k^2},$$

implying that

$$\frac{t_{k+1}}{t_k} = \frac{3 - t_k^2}{1 - 3t_k^2} = \frac{\sqrt{3} + t_k}{1 - \sqrt{3}t_k} \cdot \frac{\sqrt{3} - t_k}{1 + \sqrt{3}t_k} = u_k v_k.$$

Consequently,

$$\prod_{k=1}^{n}(u_k v_k) = \frac{t_2}{t_1} \cdot \frac{t_3}{t_2} \cdots \frac{t_{n+1}}{t_n} = \frac{\tan\left(\pi + \frac{\pi}{3^n - 1}\right)}{\tan\left(\frac{\pi}{3^n - 1}\right)} = 1,$$

establishing equation $(*)$.

24. [China 1999, by Yuming Huang] Let $P_2(x) = x^2 - 2$. Find all sequences of polynomials $\{P_k(x)\}_{k=1}^{\infty}$ such that $P_k(x)$ is monic (that is, with leading coefficient 1), has degree k, and $P_i(P_j(x)) = P_j(P_i(x))$ for all positive integers i and j.

Solution: First, we show that the sequence, if it exists, is unique. In fact, for each n, there can be only one P_n that satisfies $P_n(P_2(x)) = P_2(P_n(x))$. Let

$$P_n(x) = x^n + a_{n-1}x^{n-1} + \cdots + a_1 x + a_0.$$

By assumption,

$$\begin{aligned}(x^2 - 2)^n + a_{n-1}(x^2 - 2)^{n-1} + \cdots + a_1(x^2 - 2) + a_0 \\ = (x^n + a_{n-1}x^{n-1} + \cdots + a_1 x + a_0)^2 - 2.\end{aligned}$$

Consider the coefficients on both sides. On the left side, $2i$ is the highest power of x in which a_i appears. On the right side, the highest power is x^{n+i}, and there it appears as $2a_i x^{n+i}$. Thus, we see that the maximal power of a_i always is higher on the right side. It follows that we can solve for each a_i in turn, from $n - 1$ to 0, by equating coefficients. Furthermore, we are guaranteed that the polynomial is unique, since the equation we need to solve to find each a_i is linear.

Second, we define P_n explicitly. We claim that $P_n(x) = 2T_n\left(\frac{x}{2}\right)$ (where T_n is the nth **Chebyshev polynomial** defined in Introductory Problem 49). That is, P_n is defined by the recursive relation $P_1(x) = x$, $P_2(x) = x^2 - 2$, and

$$P_{n+1}(x) = xP_n(x) - P_{n-1}(x).$$

Because $T_n(\cos\theta) = \cos n\theta$, $P_n(2\cos\theta) = 2\cos n\theta$. It follows that for all θ,

$$P_m(P_n(2\cos\theta)) = P_m(2\cos n\theta) = 2\cos mn\theta$$
$$= P_n(2\cos m\theta) = P_n(P_m(2\cos\theta)).$$

Thus, $P_m(P_n(x))$ and $P_n(P_m(x))$ agree at all values of x in the interval $[-2, 2]$. Because both are polynomials, it follows that they are equal for all x, which completes the proof.

25. [China 2000, by Xuanguo Huang] In triangle ABC, $a \le b \le c$. As a function of angle C, determine the conditions under which $a + b - 2R - 2r$ is positive, negative, or zero.

Solution: In Figure 5.6, set $\angle A = 2x$, $\angle B = 2y$, $\angle C = 2z$. Then $0 < x \le y \le z$ and $x + y + z = \frac{\pi}{2}$. Let s denote the given quantity $a + b - 2R - 2r$. Using the **extended law of sines** and by Introductory Problem 25(d), we have

$$s = 2R(\sin 2x + \sin 2y - 1 - 4\sin x \sin y \sin z).$$

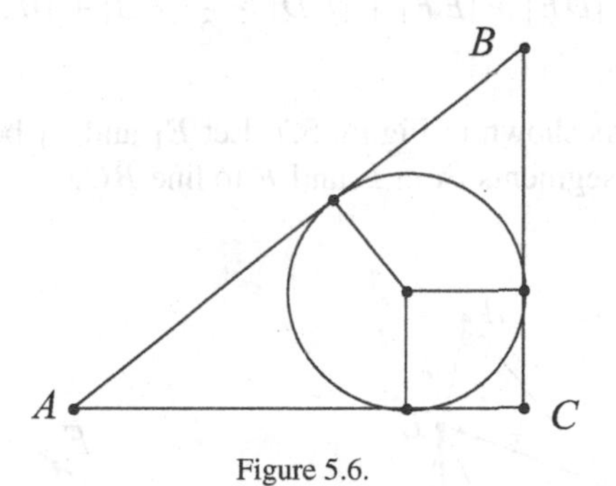

Figure 5.6.

Note that in a right triangle ABC with $\angle C = \frac{\pi}{2}$, we have $2R = c$ and $2r = a + b - c$, implying that $s = 0$. Hence, we factor $\cos 2z$ from our expression for s. By the **sum-to-product**, **product-to-sum**, and **double-angle**

formulas, we have

$$\begin{aligned}
\frac{s}{2R} &= 2\sin(x+y)\cos(x-y) - 1 + 2(\cos(x+y) - \cos(x-y))\sin z \\
&= 2\cos z\cos(x-y) - 1 + 2(\sin z - \cos(x-y))\sin z \\
&= 2\cos(x-y)(\cos z - \sin z) - \cos 2z \\
&= 2\cos(y-x)\cdot\frac{\cos^2 z - \sin^2 z}{\cos z + \sin z} - \cos 2z \\
&= \left[\frac{2\cos(y-x)}{\cos z + \sin z} - 1\right]\cos 2z,
\end{aligned}$$

where we may safely introduce the quantity $\cos z + \sin z$ because it is positive when $0 < z < \frac{\pi}{2}$.

Observe that $0 \le y - x < \min\{y, x+y\} \le \min\left\{z, \frac{\pi}{2} - z\right\}$. Because $z \le \frac{\pi}{2}$ and $\frac{\pi}{2} - z \le \frac{\pi}{2}$, we have $\cos(y-x) > \max\left\{\cos z, \cos\left(\frac{\pi}{2} - z\right)\right\} = \max\{\cos z, \sin z\}$. Hence $2\cos(y-x) > \cos z + \sin z$, or

$$\frac{2\cos(x-y)}{\cos z + \sin z} - 1 > 0.$$

Thus, $s = p\cos 2z$ for some $p > 0$. It follows that $s = a + b - 2R - 2r$ is positive, zero, or negative if and only if angle C is acute, right, or obtuse, respectively.

26. Let ABC be a triangle. Points D, E, F are on sides BC, CA, AB, respectively, such that $|DC| + |CE| = |EA| + |AF| = |FB| + |BD|$. Prove that

$$|DE| + |EF| + |FD| \ge \frac{1}{2}(|AB| + |BC| + |CA|).$$

Solution: As shown in Figure 5.7, Let E_1 and F_1 be the feet of the perpendicular line segments from E and F to line BC.

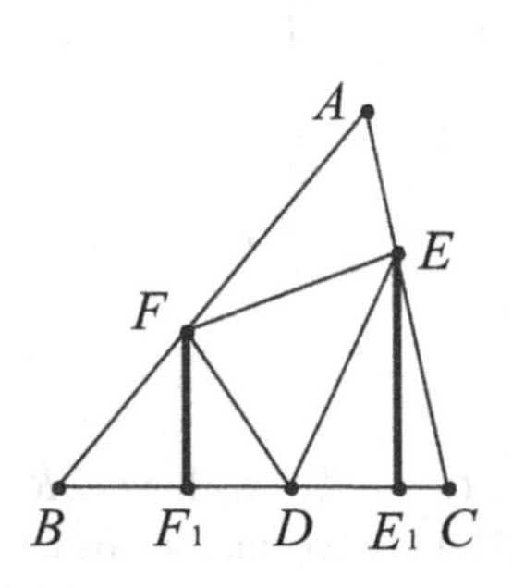

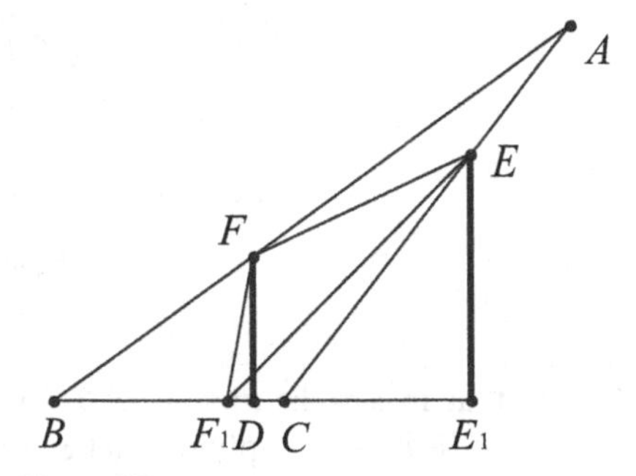

Figure 5.7.

We have

$$|EF| \geq |E_1F_1| = a - (|BF|\cos B + |CE|\cos C).$$

Likewise, we have

$$|DE| \geq c - (|AE|\cos A + |BD|\cos B)$$

and

$$|FD| \geq b - (|CD|\cos C + |AF|\cos A).$$

Note that $|DC| + |CE| = |EA| + |AF| = |FB| + |BD| = \frac{1}{3}(a + b + c)$. Adding the last three inequalities gives

$$\begin{aligned}
&|DE| + |EF| + |FD| \\
&\geq a + b + c - \frac{1}{3}(a + b + c)(\cos A + \cos B + \cos C) \\
&\geq \frac{1}{2}(a + b + c),
\end{aligned}$$

by Introductory Problem 27(b). Equality holds if and only if the length of segment EF (FD and DE) is equal to the length of the projection of segment EF on line BC (FD on CA and DE on AB), and $A = B = C = 60^\circ$, that is, if and only if D, E, and F are the midpoints of an equilateral triangle.

27. Let a and b be positive real numbers. Prove that

$$\frac{1}{\sqrt{1+a^2}} + \frac{1}{\sqrt{1+b^2}} \geq \frac{2}{\sqrt{1+ab}}$$

if either (1) $0 < a, b \leq 1$ or (2) $ab \geq 3$.

Note: Part (1) appeared in the Russian Mathematics Olympiad in 2001.

Solution: Because a and b are positive real numbers, there are angles x and y, with $0 < x, y < 90^\circ$, such that $\tan x = a$ and $\tan y = b$. The desired inequality is clearly true when $a = b$. Hence we assume that $a \neq b$, or equivalently, $x \neq y$. Then $1 + a^2 = \sec^2 x$ and $\frac{1}{\sqrt{1+a^2}} = \cos x$. Note that

$$1 + ab = \frac{\cos x \cos y + \sin x \sin y}{\cos x \cos y} = \frac{\cos(x - y)}{\cos x \cos y}$$

by the **addition and subtraction formulas**. The desired inequality reduces to

$$\cos x + \cos y \geq 2\sqrt{\frac{\cos x \cos y}{\cos(x - y)}}. \qquad (*)$$

To establish part (1), we rewrite inequality (∗) as

$$\cos^2 x + \cos^2 y + 2\cos x \cos y \le \frac{4\cos x \cos y}{\cos(x - y)}.$$

Because $0 < |x - y| < 90°$, it follows that $0 < \cos(x - y) < 1$. Hence $2\cos x \cos y \le \frac{2\cos x \cos y}{\cos(x-y)}$. It suffices to show that

$$\cos(x - y)\left[\cos^2 x + \cos^2 y\right] \le 2\cos x \cos y,$$

or

$$\cos(x - y)\,[\cos 2x + \cos 2y + 2] \le 4\cos x \cos y$$

by the **double-angle formulas**. By the **sum-to-product formulas**, the last inequality is equivalent to

$$\cos(x - y)[2\cos(x - y)\cos(x + y) + 2] \le 2[\cos(x - y) + \cos(x + y)],$$

or $\cos^2(x - y)\cos(x + y) \le \cos(x + y)$, which is clearly true, because for $0 < a, b \le 1$, we have $0° < x, y < 45°$, and so $0° < x + y \le 90°$ and $\cos(x + y) > 0$. This completes the proof of part (1).

To prove part (2), we rewrite inequality (∗) as

$$2\cos\frac{x + y}{2}\cos\frac{x - y}{2} \ge 2\sqrt{\frac{\frac{1}{2}[\cos(x + y) + \cos(x - y)]}{\cos(x - y)}}$$

by the sum-to-product and **product-to-sum formulas**. Squaring both sides of the inequality and clearing denominators gives

$$4\cos^2\frac{x + y}{2}\cos^2\frac{x - y}{2}\cos(x - y) \ge 2[\cos(x + y) + \cos(x - y)],$$

or

$$[1 + \cos(x + y)][1 + \cos(x - y)]\cos(x - y) \ge 2[\cos(x + y) + \cos(x - y)]$$

by the double-angle formulas. Setting $s = \cos(x + y)$ and $t = \cos(x - y)$, it suffices to prove that

$$(1 + s)(1 + t)t \ge 2(s + t),$$

or,

$$0 \le (1 + s)t^2 + (s - 1)t - 2s = (t - 1)[(1 + s)t + 2s].$$

Because $t \le 1$, it suffices to show that

$$(1 + s)t + 2s \le 0.$$

Because $ab \geq 3$, $\tan x \tan y \geq 3$, or equivalently, $\sin x \sin y \geq 3\cos x \cos y$. By the **product-to-sum formulas**, we have

$$\frac{1}{2}[\cos(x-y) - \cos(x+y)] \geq \frac{3}{2}[\cos(x-y) + \cos(x+y)],$$

or $t \leq -2s$. Because $1 + s \geq 0$, $(1+s)t \leq -(1+s)2s$. Consequently, $(1+s)t + 2s \leq -(1+s)2s + 2s = -2s^2 \leq 0$, as desired.

28. [China 1998, by Xuanguo Huang] Let ABC be a nonobtuse triangle such that $AB > AC$ and $\angle B = 45°$. Let O and I denote the circumcenter and incenter of triangle ABC, respectively. Suppose that $\sqrt{2}|OI| = |AB| - |AC|$. Determine all the possible values of $\sin A$.

First Solution: Applying the **extended law of sines** to triangle ABC yields $a = 2R\sin A$, $b = 2R\sin B$, and $c = 2R\sin C$. If incircle is tangent to side AB at D (Figure 5.8). Then $|BD| = \frac{c+a-b}{2}$, and so $r = |ID| = |BD|\tan\frac{B}{2}$. The **half-angle formulas** give

$$\tan\frac{B}{2} = \frac{1-\cos B}{\sin B} = \frac{1-\frac{\sqrt{2}}{2}}{\frac{\sqrt{2}}{2}} = \sqrt{2} - 1,$$

and so

$$r = R\left(\sqrt{2}-1\right)(\sin A + \sin C - \sin B).$$

By **Euler's formula**, $|OI|^2 = R(R - 2r)$, so we have

$$|OI|^2 = R^2 - 2Rr = R^2\left[1 - 2(\sin A + \sin C - \sin B)\left(\sqrt{2}-1\right)\right].$$

Squaring both sides of the given equation $\sqrt{2}|OI| = |AB| - |AC|$ gives

$$|OI|^2 = \frac{(c-b)^2}{2} = 2R^2(\sin C - \sin B)^2.$$

Therefore

$$2(\sin C - \sin B)^2 = 1 - 2(\sin A + \sin C - \sin B)\left(\sqrt{2}-1\right)$$

or

$$1 - 2\left(\sin C - \frac{\sqrt{2}}{2}\right)^2 = 2\left(\sin A + \sin C - \frac{\sqrt{2}}{2}\right)\left(\sqrt{2}-1\right). \quad (*)$$

The **addition and subtraction formulas** give

$$\begin{aligned}\sin C &= \sin(180^\circ - B - A) = \sin(135^\circ - A)\\ &= \sin 135^\circ \cos A - \cos 135^\circ \sin A = \frac{\sqrt{2}(\sin A + \cos A)}{2},\end{aligned}$$

and so

$$\sin C - \frac{\sqrt{2}}{2} = \frac{\sqrt{2}}{2}(\sin A + \cos A - 1).$$

Plugging the last equation into equation $(*)$ yields

$$1 - (\sin A + \cos A - 1)^2 = 2\left(\sqrt{2} - 1\right)\left[\sin A + \frac{\sqrt{2}}{2}(\sin A + \cos A - 1)\right].$$

Expanding both sides of the last equation gives

$$\begin{aligned}&1 - (\sin A + \cos A)^2 + 2(\sin A + \cos A) - 1\\ &= \left(\sqrt{2} - 1\right)\left(2 + \sqrt{2}\right)\sin A + \left(2 - \sqrt{2}\right)(\cos A - 1)\end{aligned}$$

or

$$\sin^2 A + \cos^2 A + 2\sin A\cos A = \left(2 - \sqrt{2}\right)\sin A + \sqrt{2}\cos A + \left(2 - \sqrt{2}\right).$$

Consequently, we have

$$2\sin A\cos A - \left(2 - \sqrt{2}\right)\sin A - \sqrt{2}\cos A + \left(\sqrt{2} - 1\right) = 0;$$

that is,

$$\left(\sqrt{2}\sin A - 1\right)\left(\sqrt{2}\cos A - \sqrt{2} + 1\right) = 0.$$

This implies that $\sin A = \frac{\sqrt{2}}{2}$ or $\cos A = 1 - \frac{\sqrt{2}}{2}$. Therefore, the answer to the problem is

$$\sin A = \frac{\sqrt{2}}{2} \quad \text{or} \quad \sin A = \sqrt{1 - \cos^2 A} = \frac{\sqrt{4\sqrt{2} - 2}}{2}.$$

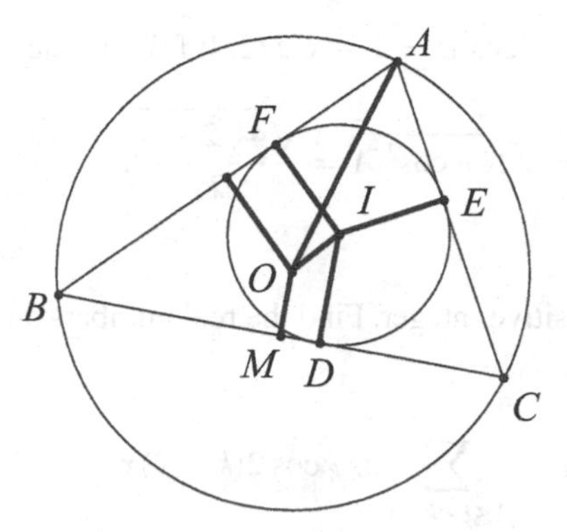

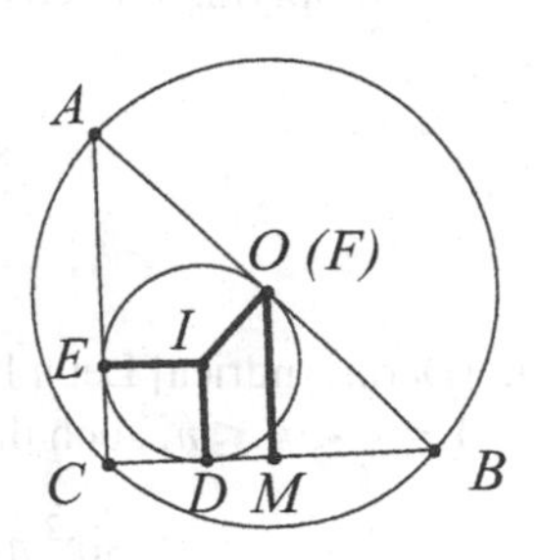

Figure 5.8.

Second Solution: As shown in Figure 5.8, the incircle touches the sides AB, BC, and CA at D, E, and F, respectively. Let M be the foot of the perpendicular line segments from O to side BC. Then the line OM is the perpendicular bisector of BC, and $|BM| = |CM|$. From equal tangents, we have $|AF| = |AE|$, $|BD| = |BF|$, and $|CD| = |CE|$. Because $c > b$, M lies on segment BD. We find that

$$\begin{aligned}\sqrt{2}|OI| &= c - b = (|AF| + |FB|) - (|AE| + |EC|)\\ &= |FB| - |EC| = |BD| - |DC|.\end{aligned}$$

We deduce that $|BD| = |BM| + |MD|$ and $|DC| = |CM| - |DM|$. Hence $\sqrt{2}|OI| = 2|DM|$, or $|OI| = \sqrt{2}|DM|$. Thus lines OI and DM form a 45° angle, which implies that either $OI \perp AB$ or $OI \parallel AB$. We consider these two cases separately.

- **First Case**: In this case, we assume that $OI \perp AB$. Then OI is the perpendicular bisector of side AB; that is, the incenter lies on the perpendicular bisector of side AB. Thus triangle ABC must be isosceles, with $|AC| = |BC|$, and so $A = B = 45^\circ$ and $\sin A = \frac{\sqrt{2}}{2}$.
- **Second Case**: In this case, we assume that $OI \parallel AB$. Let N be the midpoint of side AB. Then $OIFN$ is a rectangle. Note that $\angle AON = \angle C$, and thus

$$R \cos \angle AON = R \cos C = |ON| = |IF| = r.$$

By the solution to Introductory Problem 27, we have

$$\cos C = \frac{R}{r} = \cos A + \cos B + \cos C - 1,$$

implying that $\cos A = 1 - \cos B = 1 - \sqrt{2}/2$. It follows that

$$\sin A = \sqrt{1 - \cos^2 A} = \frac{\sqrt{4\sqrt{2} - 2}}{2}.$$

29. [Dorin Andrica] Let n be a positive integer. Find the real numbers a_0 and $a_{k,\ell}$, $1 \le \ell < k \le n$, such that

$$\frac{\sin^2 nx}{\sin^2 x} = a_0 + \sum_{1 \le \ell < k \le n} a_{\ell,k} \cos 2(k - \ell)x$$

for all real numbers x with x not an integer multiple of π.

Solution: In this solution, we apply a similar technique to that shown in the first solution of Advanced Problem 18. Note that

$$2 \sin 2kx \sin x = \cos(2k - 1)x - \cos(2k + 1)x.$$

We have

$$\begin{aligned} &2 \sin x(\sin 2x + \sin 4x + \cdots + \sin 2nx) \\ &\quad = [\cos x - \cos 3x] + [\cos 3x - \cos 5x] \\ &\qquad + \cdots + [\cos(2n - 1)x - \cos(2n + 1)x] \\ &\quad = \cos x - \cos(2n + 1)x = 2 \sin nx \sin(n + 1)x, \end{aligned}$$

or

$$s = \sin 2x + \sin 4x + \cdots + \sin 2nx = \frac{\sin nx \sin(n + 1)x}{\sin x}.$$

Similarly, by noting that

$$2 \cos 2kx \sin x = \sin(2k + 1)x - \sin(2k - 1)x,$$

we have

$$\begin{aligned} &2 \sin x(\cos 2x + \cos 4x + \cdots + \cos 2nx) \\ &\quad = [\sin 3x - \sin x] + [\sin 5x - \sin 3x] \\ &\qquad + \cdots + [\sin(2n + 1)x - \sin(2n - 1)x] \\ &\quad = \sin(2n + 1)x - \sin x = 2 \sin nx \cos(n + 1)x, \end{aligned}$$

or

$$c = \cos 2x + \cos 4x + \cdots + \cos 2nx = \frac{\sin nx \cos(n + 1)x}{\sin x}.$$

It follows that

$$\left(\frac{\sin^2 nx}{\sin^2 x}\right)^2 = \left(\frac{\sin nx \sin(n+1)x}{\sin x}\right)^2 + \left(\frac{\sin nx \cos(n+1)x}{\sin x}\right)^2$$
$$= s^2 + c^2.$$

On the other hand,

$$\begin{aligned} s^2 + c^2 &= (\sin 2x + \sin 4x + \cdots + \sin 2nx)^2 \\ &\quad + (\cos 2x + \cos 4x + \cdots + \cos 2nx)^2 \\ &= n + \sum_{1 \le \ell < k \le n} (2\sin 2\ell x \sin 2kx + 2\cos 2\ell x \cos 2kx) \\ &= n + 2 \sum_{1 \le \ell < k \le n} \cos 2(k-\ell)x \end{aligned}$$

by the **product-to-sum formulas**. Setting $a_0 = n$ and $a_{\ell,k} = 2$ solves the problem.

30. [USAMO 2000] Let S be the set of all triangles ABC for which

$$5\left(\frac{1}{|AP|} + \frac{1}{|BQ|} + \frac{1}{|CR|}\right) - \frac{3}{\min\{|AP|, |BQ|, |CR|\}} = \frac{6}{r},$$

where r is the inradius and P, Q, and R are the points of tangency of the incircle with sides AB, BC, and CA, respectively. Prove that all triangles in S are isosceles and similar to one another.

Solution: Let I be the incenter of triangle ABC. Then $|IP| = |IQ| = |IR| = r$. By symmetry, we may assume that $\min\{|AP|, |BQ|, |CR|\} = |AP|$, as shown in Figure 5.9. Let $x = \tan\frac{A}{2}$, $y = \tan\frac{B}{2}$, and $z = \tan\frac{C}{2}$. By Introductory Problem 19(a), we also have

$$xy + yz + zx = 1. \qquad (*)$$

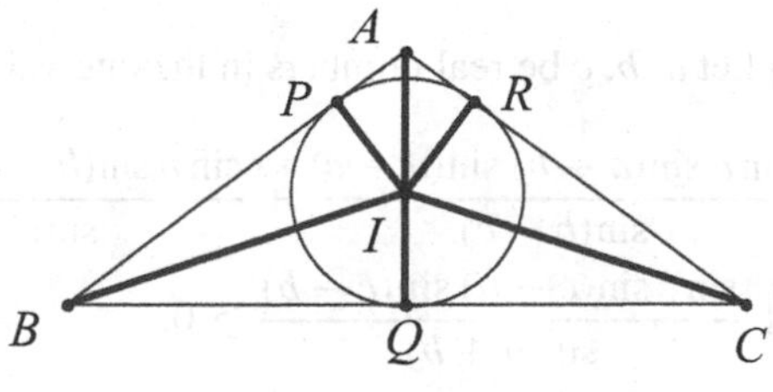

Figure 5.9.

Note that $|AP| = \frac{r}{x}$, $|BQ| = \frac{r}{y}$, and $|CR| = \frac{r}{z}$. Then the equation given in the problem statement becomes

$$2x + 5y + 5z = 6. \qquad (**)$$

Eliminating x from equations $(*)$ and $(**)$ yields

$$5y^2 + 5z^2 + 8yz - 6y - 6z + 2 = 0.$$

Completing the squares, we obtain

$$(3y - 1)^2 + (3z - 1)^2 = 4(y - z)^2.$$

Setting $3y - 1 = u$ and $3z - 1 = v$ gives $y = \frac{u+1}{3}$ and $z = \frac{v+1}{3}$, and so $y - z = \frac{u-v}{3}$. The above equation becomes

$$5u^2 + 8uv + 5v^2 = 0.$$

Because the discriminant of this quadratic equation is $8^2 - 4 \cdot 25 < 0$, the only real solution to the equation is $u = v = 0$. Thus there is only one possible set of values for the tangents of half-angles of ABC (namely, $x = \frac{4}{3}$ and $y = z = \frac{1}{3}$). Thus all triangles in S are isosceles and similar to one another.

Indeed, we have $x = \frac{r}{|AP|} = \frac{4}{3}$ and $y = z = \frac{r}{|BQ|} = \frac{r}{|CQ|} = \frac{1}{3} = \frac{4}{12}$, so we can set $r = 4$, $|AP| = |AR| = 3$, and $|BP| = |BQ| = |CQ| = |CR| = 12$. This leads to $|AB| = |AC| = 15$ and $|BC| = 24$. By scaling, all triangles in S are similar to the triangle with side lengths 5, 5, 8.

We can also use the **half-angle formulas** to calculate

$$\sin B = \sin C = \frac{2\tan\frac{C}{2}}{1 + \tan^2\frac{C}{2}} = \frac{3}{5}.$$

From this it follows that $|AQ| : |QB| : |BA| = 3 : 4 : 5$ and $|AB| : |AC| : |BC| = 5 : 5 : 8$.

31. [TST 2003] Let a, b, c be real numbers in the interval $(0, \frac{\pi}{2})$. Prove that

$$\frac{\sin a \sin(a - b)\sin(a - c)}{\sin(b + c)} + \frac{\sin b \sin(b - c)\sin(b - a)}{\sin(c + a)} + \frac{\sin c \sin(c - a)\sin(c - b)}{\sin(a + b)} \geq 0.$$

Solution: By the **product-to-sum** and the **double-angle formulas**, we have

$$\sin(\alpha-\beta)\sin(\alpha+\beta) = \frac{1}{2}[\cos 2\beta - \cos 2\alpha]$$
$$= \sin^2\alpha - \sin^2\beta.$$

Hence, we obtain

$$\sin a \sin(a-b)\sin(a-c)\sin(a+b)\sin(a+c)$$
$$= \sin a(\sin^2 a - \sin^2 b)(\sin^2 a - \sin^2 c)$$

and its analogous cyclic symmetric forms. Therefore, it suffices to prove that

$$x\left(x^2-y^2\right)\left(x^2-z^2\right)+y\left(y^2-z^2\right)\left(y^2-x^2\right)+z\left(z^2-x^2\right)\left(z^2-y^2\right)\geq 0,$$

where $x = \sin a$, $y = \sin b$, and $z = \sin c$ (hence $x, y, z > 0$). Since the last inequality is symmetric with respect to x, y, and z, we may assume that $0 < x \leq y \leq z$. It suffices to prove that

$$x\left(y^2-x^2\right)\left(z^2-x^2\right)+z\left(z^2-x^2\right)\left(z^2-y^2\right)\geq y\left(z^2-y^2\right)\left(y^2-x^2\right),$$

which is evident, because

$$x\left(y^2-x^2\right)\left(z^2-x^2\right)\geq 0$$

and

$$z\left(z^2-x^2\right)\left(z^2-y^2\right)\geq z\left(y^2-x^2\right)\left(z^2-y^2\right)\geq y\left(z^2-y^2\right)\left(y^2-x^2\right).$$

Note: The key step of the proof is an instance of **Schur's inequality** with $r = \frac{1}{2}$.

32. [TST 2002] Let ABC be a triangle. Prove that

$$\sin\frac{3A}{2}+\sin\frac{3B}{2}+\sin\frac{3C}{2}\leq\cos\frac{A-B}{2}+\cos\frac{B-C}{2}+\cos\frac{C-A}{2}.$$

First Solution: Let $\alpha = \frac{A}{2}$, $\beta = \frac{B}{2}$, $\gamma = \frac{C}{2}$. Then $0^\circ < \alpha, \beta, \gamma < 90^\circ$ and

$\alpha + \beta + \gamma = 90°$. By the **difference-to-product formulas**, we have

$$\begin{aligned}\sin\frac{3A}{2} - \cos\frac{B-C}{2} &= \sin 3\alpha - \cos(\beta - \gamma)\\ &= \sin 3\alpha - \sin(\alpha + 2\gamma)\\ &= 2\cos(2\alpha + \gamma)\sin(\alpha - \gamma)\\ &= -2\sin(\alpha - \beta)\sin(\alpha - \gamma).\end{aligned}$$

In exactly the same way, we can show that

$$\sin\frac{3B}{2} - \cos\frac{C-A}{2} = -2\sin(\beta - \alpha)\sin(\beta - \gamma)$$

and

$$\sin\frac{3C}{2} - \cos\frac{A-B}{2} = -2\sin(\gamma - \alpha)\sin(\gamma - \beta).$$

Hence it suffices to prove that

$$\sin(\alpha - \beta)\sin(\alpha - \gamma) + \sin(\beta - \alpha)\sin(\beta - \gamma) + \sin(\gamma - \alpha)\sin(\gamma - \beta) \ge 0.$$

Note that this inequality is symmetric with respect to α, β, and γ, so we can assume without loss of generality that $0° < \alpha \le \beta \le \gamma < 90°$. Then regrouping the terms on the left-hand side gives

$$\sin(\alpha - \beta)\sin(\alpha - \gamma) + \sin(\gamma - \beta)[\sin(\gamma - \alpha) - \sin(\beta - \alpha)],$$

which is positive because the function $y = \sin x$ is increasing for $0° < x < 90°$.

Note: Again the proof is similar to that of **Schur's inequality**.

Second Solution: We maintain the same notation as in the first solution. By the **addition and subtraction formulas**, we have

$$\begin{aligned}\sin 3\alpha &= \sin\alpha\cos 2\alpha + \sin 2\alpha\cos\alpha;\\ \cos(\beta - \alpha) &= \sin(2\alpha + \gamma) = \sin 2\alpha\cos\gamma + \sin\gamma\cos 2\alpha;\\ \cos(\beta - \gamma) &= \sin(2\gamma + \alpha) = \sin 2\gamma\cos\alpha + \sin\alpha\cos 2\gamma;\\ \sin 3\gamma &= \sin\gamma\cos 2\gamma + \sin 2\gamma\cos\gamma.\end{aligned}$$

By the difference-to-product formulas, it follows that

$$\begin{aligned}
&\sin 3\alpha + \sin 3\gamma - \cos(\beta - \alpha) - \cos(\beta - \gamma) \\
&\quad = (\sin\alpha - \sin\gamma)(\cos 2\alpha - \cos 2\gamma) \\
&\qquad + (\cos\alpha - \cos\gamma)(\sin 2\alpha - \sin 2\gamma) \\
&\quad = (\sin\alpha - \sin\gamma)(\cos 2\alpha - \cos 2\gamma) \\
&\qquad + 2(\cos\alpha - \cos\gamma)\cos(\alpha + \gamma)\sin(\alpha - \gamma).
\end{aligned}$$

Note that $\sin x$ is increasing, and $\cos x$ and $\cos 2x$ are decreasing for $0^\circ < x < 90^\circ$. Since $0^\circ < \alpha, \gamma, \alpha + \gamma < 90^\circ$, each of the two products in the last sum is less than or equal to 0. Hence

$$\sin 3\alpha + \sin 3\gamma - \cos(\beta - \alpha) - \cos(\beta - \gamma) \le 0.$$

In exactly the same way, we can show that

$$\sin 3\beta + \sin 3\alpha - \cos(\gamma - \beta) - \cos(\gamma - \alpha) \le 0$$

and

$$\sin 3\gamma + \sin 3\beta - \cos(\alpha - \gamma) - \cos(\alpha - \beta) \le 0.$$

Adding the last three inequalities gives the desired result.

33. Let $x_1, x_2, \ldots, x_n$, $n \ge 2$, be n distinct real numbers in the interval $[-1, 1]$. Prove that

$$\frac{1}{t_1} + \frac{1}{t_2} + \cdots + \frac{1}{t_n} \ge 2^{n-2},$$

where $t_i = \prod_{j \ne i} |x_j - x_i|$.

Solution: Let T_n denote the nth **Chebyshev polynomial**. Recall that (Introductory Problem 49) $T_n(\cos x) = \cos nx$ and T_n is defined by the recursion $T_{n+1}(x) = 2xT_n(x) - T_{n-1}(x)$, $T_0(x) = 1$, and $T_1(x) = x$. Therefore, the leading coefficient of T_n is 2^{n-1} for $n \ge 1$.

Now we apply the above information to the problem at hand. We can apply **Lagrange's interpolation formula** to the points $x_1, x_2, \ldots, x_n$ and the polynomial $T_{n-1}(x)$ to obtain

$$T_{n-1}(x) = \sum_{k=1}^{n} \frac{T_{n-1}(x_k)(x - x_1)\cdots(x - x_{k-1})(x - x_{k+1})\cdots(x - x_n)}{(x_k - x_1)\cdots(x_k - x_{k-1})(x_k - x_{k+1})\cdots(x_k - x_n)}.$$

Equating leading coefficients, we have

$$2^{n-2} = \sum_{k=1}^{n} \frac{T_{n-1}(x_k)}{(x_k - x_1)\cdots(x_k - x_{k-1})(x_k - x_{k+1})\cdots(x_k - x_n)}.$$

Set θ_k such that $\cos\theta_k = x_k$. Then $|T_{n-1}(x_k)| = |\cos(n-1)\theta_k| \le 1$. It follows that

$$\begin{aligned} 2^{n-2} &\le \sum_{k=1}^{n} \frac{|T_{n-1}(x_k)|}{|(x_k - x_1)\cdots(x_k - x_{k-1})(x_k - x_{k+1})\cdots(x_k - x_n)|} \\ &= \sum_{k=1}^{n} \frac{1}{t_k}, \end{aligned}$$

as desired.

34. [St. Petersburg 2001] Let $x_1, \ldots, x_{10}$ be real numbers in the interval $[0, \pi/2]$ such that $\sin^2 x_1 + \sin^2 x_2 + \cdots + \sin^2 x_{10} = 1$. Prove that

$$3(\sin x_1 + \cdots + \sin x_{10}) \le \cos x_1 + \cdots + \cos x_{10}.$$

Solution: Because $\sin^2 x_1 + \sin^2 x_2 + \cdots + \sin^2 x_{10} = 1$,

$$\cos x_i = \sqrt{\sum_{j \ne i} \sin^2 x_j}.$$

By the **power mean inequality**, for each $1 \le i \le 10$,

$$\cos x_i = \sqrt{\sum_{j \ne i} \sin^2 x_j} \ge \frac{\sum_{j \ne i} \sin x_j}{3}.$$

Summing over all the terms $\cos x_i$ gives

$$\sum_{i=1}^{10} \cos x_i \ge \sum_{i=1}^{10} \sum_{j \ne i} \frac{\sin x_j}{3} = \sum_{i=1}^{10} 9 \cdot \frac{\sin x_i}{3} = 3 \sum_{i=1}^{10} \sin x_i,$$

as desired.

35. [IMO 2001 shortlist] Let $x_1, x_2, \ldots, x_n$ be arbitrary real numbers. Prove the inequality

$$\frac{x_1}{1+x_1^2} + \frac{x_2}{1+x_1^2+x_2^2} + \cdots + \frac{x_n}{1+x_1^2+\cdots+x_n^2} < \sqrt{n}.$$

Solution: (By Ricky Liu) We make the following substitutions: $x_1 = \tan\alpha_1$, $x_2 = \sec\alpha_1 \tan\alpha_2$, and

$$x_k = \sec\alpha_1 \sec\alpha_2 \cdots \sec\alpha_{k-1} \tan\alpha_k,$$

with $-\pi/2 < \alpha_k < \pi/2$, $1 \le k \le n$. Note that this is always possible because the range of $\tan\alpha$ is $(-\infty, \infty)$ and $\sec\alpha$ is always nonzero. Then the kth term on the left-hand side of our inequality becomes

$$\frac{\sec\alpha_1 \cdots \sec\alpha_{k-1} \tan\alpha_k}{1+\tan^2\alpha_1 + \cdots + \sec^2\alpha_1 \cdots \sec^2\alpha_{n-1} \tan^2\alpha_n}$$
$$= \cos\alpha_1 \cos\alpha_2 \cdot \cos\alpha_k \sin\alpha_k.$$

Hence the given inequality reduces to

$$\cos\alpha_1 \sin\alpha_1 + \cos\alpha_1 \cos\alpha_2 \sin\alpha_2 + \cdots + \cos\alpha_1 \cos\alpha_2 \cdots \cos\alpha_n \sin\alpha_n$$
$$< \sqrt{n};$$

that is,

$$c_1 s_1 + c_1 c_2 s_2 + \cdots + c_1 c_2 \cdots c_n s_n < \sqrt{n},$$

where $c_i = \cos\alpha_i$ and $s_i = \sin\alpha_i$ for $1 \le i \le n$. For $2 \le i \le n$, because $c_i^2 + s_i^2 = \cos^2\alpha_i + \sin^2\alpha_i = 1$, we note that

$$c_1^2 c_2^2 \cdots c_{i-1}^2 s_i^2 + c_1^2 c_2^2 \cdots c_{i-1}^2 c_i^2 = c_1^2 c_2^2 \cdots c_{i-1}^2.$$

Therefore,

$$s_1^2 + c_1^2 s_2^2 + \cdots + c_1^2 c_2^2 \cdots c_{n-2}^2 s_{n-1}^2 + c_1^2 c_2^2 \cdots c_{n-1}^2 = 1. \qquad (*)$$

By $(*)$ and **Cauchy–Schwarz inequality**, we obtain

$$\begin{aligned}
&c_1 s_1 + c_1 c_2 s_2 + \cdots + c_1 c_2 \cdots c_n s_n \\
&\quad \le \sqrt{s_1^2 + c_1^2 s_2^2 + \cdots + c_1^2 c_2^2 \cdots c_{n-2}^2 s_{n-1}^2 + c_1^2 c_2^2 \cdots c_{n-1}^2} \\
&\qquad \cdot \sqrt{c_1^2 + c_2^2 + \cdots + c_{n-1}^2 + c_n^2 s_n^2} \\
&\quad = \sqrt{c_1^2 + c_2^2 + \cdots + c_{n-1}^2 + c_n^2 s_n^2} \\
&\quad = \sqrt{\cos^2\alpha_1 + \cos^2\alpha_2 + \cdots + \cos^2\alpha_{n-1} + \cos^2\alpha_n \sin^2\alpha_n} \\
&\quad \le \sqrt{n}.
\end{aligned}$$

Equality in the last step can hold only when

$$\cos \alpha_1 = \cos \alpha_2 = \cdots = \cos \alpha_{n-1} = \cos \alpha_n \sin \alpha_n = 1,$$

which is impossible, because $\cos \alpha_n \sin \alpha_n = \frac{1}{2} \sin 2\alpha_n < 1$. Therefore, we always have strict inequality, and we are done.

36. [USAMO 1998] Let $a_0, a_1, \ldots, a_n$ be numbers in the interval $\left(0, \frac{\pi}{2}\right)$ such that

$$\tan\left(a_0 - \frac{\pi}{4}\right) + \tan\left(a_1 - \frac{\pi}{4}\right) + \cdots + \tan\left(a_n - \frac{\pi}{4}\right) \geq n - 1.$$

Prove that

$$\tan a_0 \tan a_1 \cdots \tan a_n \geq n^{n+1}.$$

Solution: Let $b_k = \tan\left(a_k - \frac{\pi}{4}\right)$, $k = 0, 1, \ldots, n$. It follows from the hypothesis that for each k, $-1 < b_k < 1$, and

$$1 + b_k \geq \sum_{0 \leq \ell \neq k \leq n} (1 - b_\ell). \tag{$*$}$$

Applying the **arithmetic–geometric means inequality** to the positive real numbers $1 - b_\ell$, $\ell = 0, 1, \ldots, k-1, k+1, \ldots, n$, we obtain

$$\sum_{0 \leq \ell \neq k \leq n} (1 - b_\ell) \geq n \left(\prod_{0 \leq \ell \neq k \leq n} (1 - b_\ell) \right)^{1/n}. \tag{$**$}$$

From inequalities $(*)$ and $(**)$ it follows that

$$\prod_{k=0}^{n} (1 + b_k) \geq n^{n+1} \left(\prod_{\ell=0}^{n} (1 - b_\ell)^n \right)^{1/n},$$

and hence that

$$\prod_{k=0}^{n} \frac{1 + b_k}{1 - b_k} \geq n^{n+1}.$$

Because

$$\frac{1 + b_k}{1 - b_k} = \frac{1 + \tan\left(a_k - \frac{\pi}{4}\right)}{1 - \tan\left(a_k - \frac{\pi}{4}\right)} = \tan\left[\left(a_k - \frac{\pi}{4}\right) + \frac{\pi}{4}\right] = \tan a_k,$$

the conclusion follows.

Note: Using a similar method, one can show that

$$\frac{1}{n-1+a_1}+\frac{1}{n-1+a_2}+\cdots+\frac{1}{n-1+a_n}\le 1,$$

where $a_1, a_2, \ldots, a_n$ are positive real numbers such that $a_1a_2\cdots a_n = 1$. An interesting exercise is to provide a trigonometry interpretation for the last inequality.

37. [MOSP 2001] Find all triples of real numbers (a, b, c) such that $a^2-2b^2 = 1$, $2b^2-3c^2 = 1$, and $ab+bc+ca = 1$.

Solution: Since $a^2-2b^2 = 1$, $a \neq 0$. Since $2b^2-3c^2 = 1$, $b \neq 0$. If $c = 0$, then $b = 1/\sqrt{2}$ and $a = \sqrt{2}$. It is easy to check that $(a, b, c) = \left(\sqrt{2}, 1/\sqrt{2}, 0\right)$ is a solution of the system. We claim that there are no other valid triples.

We approach the problem indirectly by assuming that there exists a triple of real numbers (a, b, c), with $abc \neq 0$, such that the equations hold. Without loss of generality, we assume that two of the numbers are positive; otherwise, we can consider the triple $(-a, -b, -c)$. Without loss of generality, we assume that a and b are positive. (The first two equations are independent of the signs of a, b, c, and the last equation is symmetric with respect to a, b, and c.) By Introductory Problem 21, we may assume that $a = \cot A$, $b = \cot B$, and $c = \cot C$, with $0 < A, B < 90°$, where A, B, C are angles of a triangle. We have

$$a^2+1 = 2\left(b^2+1\right) = 3\left(c^2+1\right).$$

The last equation reduces to

$$\csc^2 A = 2\csc^2 B = 3\csc^2 C,$$

or

$$\frac{1}{\sin A} = \frac{\sqrt{2}}{\sin B} = \frac{\sqrt{3}}{\sin C}.$$

By the **law of sines**, we conclude that the sides opposite angles A, B, C have lengths $k, \sqrt{2}k, \sqrt{3}k$, respectively, for some positive real number k. But then triangle ABC is a right triangle with $\angle C = 90°$, implying that $c = \cot C = 0$, a contradiction to the assumption that $c \neq 0$. Hence our assumption was wrong, and $(a, b, c) = \left(\sqrt{2}, 1/\sqrt{2}, 0\right)$ is the only valid triple sought.

38. Let n be a positive integer, and let θ_i be angles with $0 < \theta_i < 90^\circ$ such that

$$\cos^2\theta_1 + \cos^2\theta_2 + \cdots + \cos^2\theta_n = 1.$$

Prove that

$$\tan\theta_1 + \tan\theta_2 + \cdots + \tan\theta_n \geq (n-1)(\cot\theta_1 + \cot\theta_2 + \cdots + \cot\theta_n).$$

Solution: (By Tiankai Liu) By the **power mean inequality**, for positive numbers $x_1, x_2, \ldots, x_n$, we have $M_{-1} \leq M_1 \leq M_2$; that is,

$$\frac{n}{\frac{1}{x_1} + \frac{1}{x_2} + \cdots + \frac{1}{x_n}} \leq \frac{x_1 + x_2 + \cdots + x_n}{n} \leq \sqrt{\frac{x_1^2 + x_2^2 + \cdots + x_n^2}{n}}.$$

For $1 \leq i \leq n$, let $\cos\theta_i = a_i$. Then

$$\begin{aligned}\tan\theta_i &= \frac{\sin\theta_i}{\cos\theta_i} = \frac{\sqrt{1-\cos^2\theta_i}}{a_i} \\ &= \frac{\sqrt{a_1^2 + a_2^2 + \cdots + a_{i-1}^2 + a_{i+1}^2 + \cdots + a_n^2}}{a_i} \\ &\geq \frac{a_1 + a_2 + \cdots + a_{i-1} + a_{i+1} + \cdots + a_n}{a_i\sqrt{n-1}}.\end{aligned}$$

Summing the above inequalities for i from 1 to n gives

$$\sum_{i=1}^{n} \tan\theta_i \geq \frac{1}{\sqrt{n-1}} \sum_{i=1}^{n} \sum_{j \neq i} \frac{a_j}{a_i} = \frac{1}{\sqrt{n-1}} \sum_{\substack{1 \leq i, j \leq n \\ i \neq j}} \frac{a_j}{a_i}, \qquad (*)$$

because each ratio $\frac{a_i}{a_j}$ appears exactly once.

On the other hand, we have

$$\begin{aligned}\cot\theta_i &= \frac{\cos\theta_i}{\sin\theta_i} = \frac{a_i}{\sqrt{1-\cos^2\theta_i}} \\ &= \frac{a_i}{\sqrt{a_1^2 + a_2^2 + \cdots + a_{i-1}^2 + a_{i+1}^2 + \cdots + a_n^2}} \\ &\leq \frac{a_i\left(\frac{1}{a_1} + \frac{1}{a_2} + \cdots + \frac{1}{a_{i-1}} + \frac{1}{a_{i+1}} + \cdots + \frac{1}{a_n}\right)}{(n-1)\sqrt{n-1}},\end{aligned}$$

by the power mean inequality. Summing the above identities from 1 to n yields

$$\sum_{i=1}^{n} \cot\theta_i \le \frac{1}{(n-1)^{3/2}} \sum_{i=1}^{n} \sum_{j\neq i} \frac{a_i}{a_j} = \frac{1}{(n-1)^{3/2}} \sum_{\substack{1\le i,j\le n\\ i\neq j}} \frac{a_i}{a_j} \qquad (**)$$

again, because each ratio $\frac{a_i}{a_j}$ appears once. Combining inequalities $(*)$ and $(**)$ gives

$$\sqrt{n-1}\sum_{i=1}^{n} \tan\theta_i \ge \sum_{\substack{1\le i,j\le n\\ i\neq j}} \frac{a_j}{a_i} = \sum_{\substack{1\le i,j\le n\\ i\neq j}} \frac{a_i}{a_j} \ge (n-1)^{3/2} \sum_{i=1}^{n} \cot\theta_i,$$

from which the desired result follows.

39. [Weichao Wu] One of the two inequalities

$$(\sin x)^{\sin x} < (\cos x)^{\cos x} \quad \text{and} \quad (\sin x)^{\sin x} > (\cos x)^{\cos x}$$

is always true for all real numbers x such that $0 < x < \frac{\pi}{4}$. Identify that inequality and prove your result.

Solution: The first inequality is true. Observe that the logarithm function is concave down. We apply **Jensen's inequality** to the points $\sin x < \cos x < \sin x + \cos x$ with weights $\lambda_1 = \tan x$ and $\lambda_2 = 1 - \tan x$ (because $0 < x < \pi/4$, λ_1 and λ_2 are positive) to obtain

$$\begin{aligned}\log(\cos x) &= \log[\tan x \sin x + (1-\tan x)(\sin x + \cos x)] \\ &> \tan x \log(\sin x) + (1-\tan x)\log(\sin x + \cos x).\end{aligned}$$

Since $\sin x + \cos x = \sqrt{2}\sin\left(x + \frac{\pi}{4}\right) > 1$ and $\tan x < 1$ in the specified interval, the second term is positive and we may drop it to obtain

$$\log(\cos x) > \tan x \log(\sin x).$$

Multiplying by $\cos x$ and exponentiating gives the required inequality.

40. Let k be a positive integer. Prove that $\sqrt{k+1} - \sqrt{k}$ is not the real part of the complex number z with $z^n = 1$ for some positive integer n.

Note: In June 2003, this problem was first given in the training of the Chinese IMO team, and then in the MOSP. The following solution was due to Anders Kaseorg, gold medalist in the 44th IMO in July 2003 in Tokyo, Japan.

Solution: Assume to the contrary that $\alpha = \sqrt{k+1} - \sqrt{k}$ is the real part of some complex number z with $z^n = 1$ for some positive integer n. Because z is an nth root of unity, it can be written as $\cos \frac{2\pi j}{n} + i \sin \frac{2\pi j}{n}$ for some integer j with $0 \le j \le n-1$. Thus, $\alpha = \cos \frac{2\pi j}{n}$.

Let $T_n(x)$ be the nth **Chebyshev polynomial**; that is, $T_0(x) = 1$, $T_1(x) = x$, and $T_{i+1} = 2xT_i(x) - T_{i-1}(x)$ for $i \ge 1$. Then $T_n(\cos\theta) = \cos(n\theta)$, implying that $T_n(\alpha) = \cos(2\pi j) = 1$.

Let $\beta = \sqrt{k+1} + \sqrt{k}$. Note that $\alpha\beta = 1$ and $\alpha + \beta = 2\sqrt{k+1}$, and so $\alpha^2 + \beta^2 = (\alpha+\beta)^2 - 2\alpha\beta = 4k+2$. Thus $\pm\alpha$ and $\pm\beta$ are the roots of the polynomial

$$\begin{aligned} P(x) &= (x-\alpha)(x+\alpha)(x-\beta)(x+\beta) = \left(x^2 - \alpha^2\right)\left(x - \beta^2\right) \\ &= x^4 - (4k+2)x + 1. \end{aligned}$$

Let $Q(x)$ be the **minimal polynomial** for α. If neither β nor $-\beta$ is a root of $Q(x)$, then $Q(x)$ must divide

$$(x-\alpha)(x+\alpha) = x^2 - [2k+1 - 2\sqrt{k(k+1)}],$$

and so $k(k+1)$ must be a perfect square, which is impossible because $k^2 < k(k+1) < (k+1)^2$. Therefore, either $Q(\beta)$ or $Q(-\beta) = 0$ or both. We say that $Q(\beta') = 0$, with either $\beta' = \beta$ or $\beta' = -\beta$.

Because α is a root of $T_n(x) - 1$, $Q(x)$ divides $T_n(x) - 1$ and β' is a root of $T_n(x) - 1$. However, by Introductory Problem 49(f), the roots of $T_n(x) - 1$ are all in the interval $[-1, 1]$ and $|\beta'| = \sqrt{k+1} + \sqrt{k} > 1$, which is a contradiction. Therefore, our original assumption was wrong, and $\sqrt{k+1} - \sqrt{k}$ is not the real part of any nth root of unity.

41. Let $A_1A_2A_3$ be an acute-angled triangle. Points B_1, B_2, B_3 are on sides A_2A_3, A_3A_1, A_1A_2, respectively. Prove that

$$2(b_1 \cos A_1 + b_2 \cos A_2 + b_3 \cos A_3) \ge a_1 \cos A_1 + a_2 \cos A_2 + a_3 \cos A_3,$$

where $a_i = |A_{i+1}A_{i+2}|$ and $b_i = |B_{i+1}B_{i+2}|$ for $i = 1, 2, 3$ (with indices taken modulo 3; that is, $x_{i+3} = x_i$).

Solution: As shown in Figure 5.10, let $|B_iA_{i+1}| = s_i$ and $|B_iA_{i+2}| = t_i$, $i = 1, 2, 3$. Then $a_i = s_i + t_i$. Our approach here is similar to that of Advanced Problem 26. Let $A_1 = \angle A_1$, $A_2 = \angle A_2$, and $A_3 = \angle A_3$.

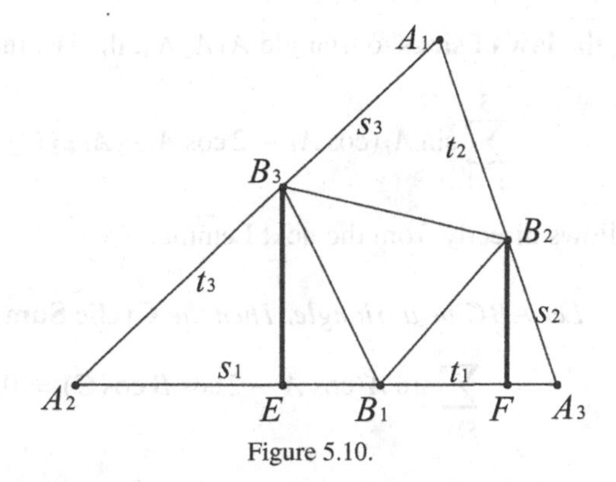

Figure 5.10.

Note that segment EF, the projection of segment B_2B_3 onto the line A_2A_3, has length $a_1 - t_3 \cos A_2 - s_2 \cos A_3$, and so

$$b_1 \geq a_1 - t_3 \cos A_2 - s_2 \cos A_3.$$

Because $0 < A_1 < 90°$, we know that

$$b_1 \cos A_1 \geq a_1 \cos A_1 - t_3 \cos A_2 \cos A_1 - s_2 \cos A_3 \cos A_1.$$

Likewise, we find that

$$b_2 \cos A_2 \geq a_2 \cos A_2 - t_1 \cos A_3 \cos A_2 - s_3 \cos A_1 \cos A_2$$

and

$$b_3 \cos A_3 \geq a_3 \cos A_3 - t_2 \cos A_1 \cos A_3 - s_1 \cos A_2 \cos A_3.$$

Adding the last three inequalities, we observe that

$$\sum_{i=1}^{3} b_i \cos A_i \geq \sum_{i=1}^{3} a_i (\cos A_i - \cos A_{i+1} A_{i+2}).$$

It suffices to show that

$$2 \sum_{i=1}^{3} a_i (\cos A_i - \cos A_{i+1} A_{i+2}) \geq \sum_{i=1}^{3} a_i \cos A_i,$$

or

$$\sum_{i=1}^{3} a_i (\cos A_i - 2 \cos A_{i+1} A_{i+2}) \geq 0.$$

Applying the **law of sines** to triangle $A_1A_2A_3$, the last inequality reduces to

$$\sum_{i=1}^{3} \sin A_i (\cos A_i - 2\cos A_{i+1}A_{i+2}) \geq 0,$$

which follows directly from the next Lemma.

Lemma *Let ABC be a triangle. Then the* **Cyclic Sum**

$$\sum_{\text{cyc}} \sin A(\cos A - 2\cos B\cos C) = 0.$$

Proof: By the **double-angle formulas**, it suffices to show that

$$\sum_{\text{cyc}} \sin 2A = 2\sum_{\text{cyc}} \sin A\cos A = 4\sum_{\text{cyc}} \sin A\cos B\cos C.$$

Applying the **addition and subtraction formulas** gives

$$\begin{aligned} & \sin A\cos B\cos C + \sin B\cos C\cos A \\ = \ & \cos C(\sin A\cos B + \sin B\cos A) \\ = \ & \cos C\sin(A+B) = \cos C\sin C. \end{aligned}$$

Hence

$$\begin{aligned} & 4\sum_{\text{cyc}} \sin A\cos B\cos C \\ = \ & 2\sum_{\text{cyc}} (\sin A\cos B\cos C + \sin B\cos C\cos A) \\ = \ & 2\sum_{\text{cyc}} \cos C\sin C = \sum_{\text{cyc}} \sin 2C, \end{aligned}$$

as desired. ■

42. Let ABC be a triangle. Let x, y, and z be real numbers, and let n be a positive integer. Prove the following four inequalities.

    (a) [D. Barrow] $x^2 + y^2 + z^2 \geq 2yz\cos A + 2zx\cos B + 2xy\cos C$.

    (b) [J. Wolstenholme]

    $$x^2 + y^2 + z^2 \geq 2(-1)^{n+1}(yz\cos nA + zx\cos nB + xy\cos nC).$$

(c) [O. Bottema] $yza^2 + zxb^2 + xyc^2 \le R^2(x + y + z)^2$.

(d) [A. Oppenheim] $xa^2 + yb^2 + zc^2 \ge 4[ABC]\sqrt{xy + yz + zx}$.

Note: These are very powerful inequalities, because x, y, z can be arbitrary real numbers. By the same token, however, they are not easy to apply.

Solution: It is clear that part (a) is a special case of part (b) by setting $n = 1$. Parts (c) and (d) are two applications of part (b). Hence we prove only parts (b), (c), and (d).

(b) Rewrite the desired inequality as

$$x^2 + 2x(-1)^n(z\cos nB + y\cos nC) + y^2 + z^2 + 2(-1)^n yz\cos nA \ge 0.$$

Completing the square for $x^2 + 2x(-1)^n(z\cos nB + y\cos C)$ gives

$$\left[x^2 + (-1)^n(z\cos nB + y\cos nC)\right]^2 + y^2 + z^2 + 2(-1)^n yz\cos nA$$
$$\ge (z\cos nB + y\cos nC)^2$$
$$= z^2\cos^2 nB + y^2\cos^2 nC + 2yz\cos nB\cos nC.$$

It suffices to show that

$$y^2 + z^2 + 2(-1)^n yz\cos nA \ge z^2\cos^2 nB + y^2\cos^2 nC + 2yz\cos nB\cos nC,$$

or

$$y^2\sin^2 nC + z^2\sin^2 nB + 2yz\left[(-1)^n\cos nA - \cos nB\cos nC\right] \ge 0. \qquad (*)$$

If $n = 2k$ is even, then $nA + nB + nC = 2k\pi$, and so $\cos nA = \cos(nB + nC) = \cos nB\cos nC - \sin nB\sin nC$. The desired inequality $(*)$ reduces to

$$y^2\sin^2 nC + z^2\sin^2 nB - 2yz\sin nB\sin nC$$
$$= (y\sin nC - z\sin nB)^2 \ge 0,$$

which is evident.

If $n = 2k + 1$ is even, then $nA + nB + nC = (2k + 1)\pi$, and so $\cos nA = -\cos(nB + nC) = -\cos nB\cos nC + \sin nB\sin nC$. The desired inequality $(*)$ reduces to

$$y^2 \sin^2 nC + z^2 \sin^2 nB - 2yz \sin nB \sin nC$$
$$= (y \sin nC - z \sin nB)^2 \geq 0,$$

which is evident.

From the above proof, we note that the equality case of the desired inequality holds only if $(y \sin nC - z \sin nB)^2 \geq 0$, that is, if $y \sin nC = z \sin nB$, or $\frac{y}{\sin nB} = \frac{z}{\sin nC}$. By symmetry, the equality case holds only if

$$\frac{x}{\sin nA} = \frac{y}{\sin nB} = \frac{z}{\sin nC}.$$

It is also easy to check that the above condition is sufficient for the equality case to hold.

(c) The **extended law of sines** gives that $\frac{a}{R} = 2 \sin A$ and its analogous forms for $\frac{b}{R}$ and $\frac{c}{R}$. Dividing both sides of the desired inequalities by R^2 and expanding the resulting right-hand side yields

$$4(yz \sin^2 A + zx \sin^2 B + xy \sin^2 C)$$
$$\leq x^2 + y^2 + z^2 + 2(xy + yz + zx),$$

or, by the **double-angle formulas**,

$$x^2 + y^2 + z^2$$
$$\geq 2\left[yz(2\sin^2 A - 1) + zx(2\sin^2 B - 1) + xy(2\sin^2 C - 1\right]$$
$$= -2(yz \cos 2A + zx \cos 2B + xy \cos 2C),$$

which is part (b) by setting $n = 2$.

By the argument at the end of the proof of (b), we conclude that the equality case of the desired inequality holds if and only if

$$\frac{x}{\sin 2A} = \frac{y}{\sin 2B} = \frac{z}{\sin 2C}.$$

(d) Setting $x = xa^2$, $y = yb^2$, and $z = zc^2$ in part (c) gives

$$a^2b^2c^2(xy + yz + zx) \leq R^2\left(xa^2 + yb^2 + zc^2\right)^2,$$

or

$$16R^2[ABC](xy + yz + zx) \leq R^2\left(xa^2 + yb^2 + zc^2\right)^2,$$

by Introductory Problem 25(a). Dividing both sides of the last inequality by R^2 and taking square roots yields the desired result.

By the argument at the end of the proof of (b), we conclude that the equality case of the desired inequality holds if and only if

$$\frac{xa^2}{\sin 2A} = \frac{yb^2}{\sin 2B} = \frac{zc^2}{\sin 2C},$$

or

$$\frac{xa}{\cos A} = \frac{yb}{\cos B} = \frac{zc}{\sin C},$$

by the double-angle formulas and the law of sines. By the **law of cosines**, we have

$$\frac{a}{\cos A} = \frac{2abc}{b^2+c^2-a^2}$$

and its analogous forms for $\frac{b}{\cos B}$ and $\frac{c}{\cos C}$. Therefore, the equality case holds if and only if

$$\frac{x}{b^2+c^2-a^2} = \frac{y}{c^2+a^2-b^2} = \frac{z}{a^2+b^2-c^2}.$$

Note: The approach of completing the squares, shown in the proof of (b), is rather tricky. We can see this approach from another angle. Consider the quadratic function

$$f(x) = x^2 - 2x(z\cos B + y\cos C) + y^2 + z^2 - 2yz\cos A.$$

Its discriminant is

$$\begin{aligned}
\Delta &= 4(z\cos B + y\cos C)^2 - 4(y^2 + z^2 - 2yz\cos A)\\
&= 4\left[z^2\cos^2 B - z^2 + 2yz(\cos A + \cos B\cos C) + y^2\cos^2 C - y^2\right]\\
&= 4\left[-z^2\sin^2 B + 2yz(-\cos(B+C) + \cos B\cos C) - y^2\sin^2 C\right]\\
&= 4\left[-z^2\sin^2 B + 2yz\sin B\sin C - y^2\sin^2 C\right]\\
&= -4(z\sin B - y\sin C)^2 \le 0.
\end{aligned}$$

For fixed y and z and large x, $f(x)$ is certainly positive. Hence $f(x) \ge 0$ for all x, establishing (a). This method can be easily generalized to prove (b). We leave the generalization to the reader as an exercise.

43. [USAMO 2004] A circle ω is inscribed in a quadrilateral $ABCD$. Let I be the center of ω. Suppose that

$$(|AI| + |DI|)^2 + (|BI| + |CI|)^2 = (|AB| + |CD|)^2.$$

Prove that $ABCD$ is an isosceles trapezoid.

Note: We introduce two trigonometric solutions and a synthetic-geometric solution. The first solution, by Oleg Golberg, is very technical. The second solution, by Tiankai Liu and Tony Zhang, reveals more geometrical background in their computations. This is by far the most challenging problem in the USAMO 2004. There were only four complete solutions. The fourth student is Jacob Tsimerman, from Canada. Evidently, these four students placed top four in the contest. There were nine IMO gold medals won by these four students, with each of Oleg and Tiankai winning three, Jacob two, and Tony one. Oleg won his first two representing Russia, and the third representing the United States. Jacob is one of only four students who achieved a perfect score at the IMO 2004 in Athens, Greece.

The key is to recognize that the given identity is a combination of equality cases of certain inequalities. By equal tangents, we have $|AB| + |CD| = |AD| + |BC|$ if and only if $ABCD$ has an incenter. We will prove that for a convex quadrilateral $ABCD$ with incenter I, then

$$(|AI|+|DI|)^2+(|BI|+|CI|)^2 \le (|AB|+|CD|)^2 = (|AD|+|BC|)^2. \quad (*)$$

Equality holds if and only if $AD \parallel BC$ and $|AB| = |CD|$. Without loss of generality, we may assume that the inradius of $ABCD$ is 1.

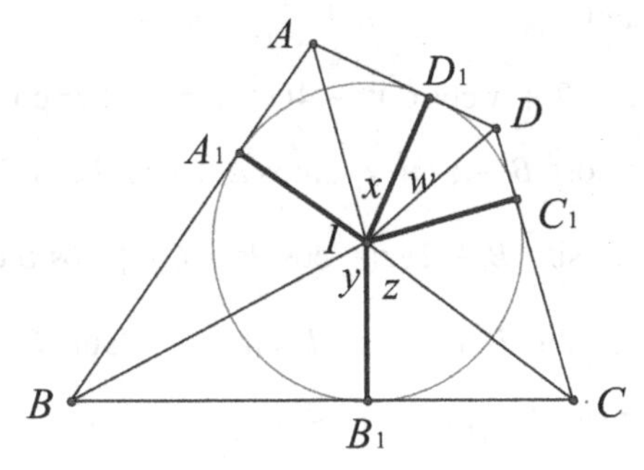

Figure 5.11.

First Solution: As shown in Figure 5.11, let A_1, B_1, C_1, and D_1 be the points of tangency. Because circle ω is inscribed in $ABCD$, we can set $\angle D_1IA = \angle AIA_1 = x$, $\angle A_1IB = \angle BIB_1 = y$, $\angle B_1IC = \angle CIC_1 = z$, $\angle C_1ID = \angle DID_1 = w$, and $x + y + z + w = 180^\circ$, or $x + w = 180^\circ - (y + z)$, with $0° < x, y, z, w < 90°$. Then $|AI| = \sec x$, $|BI| = \sec y$, $|CI| = \sec z$, $|DI| = \sec w$, $|AD| = |AD_1| + |D_1D| = \tan x + \tan w$, and $|BC| = |BB_1| + |B_1C| = \tan y + \tan z$. Inequality $(*)$ becomes

$$(\sec x + \sec w)^2 + (\sec y + \sec z)^2 \leq (\tan x + \tan y + \tan z + \tan w)^2.$$

Expanding both sides of the above inequality and applying the identity $\sec^2 x = 1 + \tan^2 x$ gives

$$\begin{aligned} &4 + 2(\sec x \sec w + \sec y \sec z) \\ &\leq 2\tan x \tan y + 2\tan x \tan z + 2\tan x \tan w \\ &\quad + 2\tan y \tan z + 2\tan y \tan w + 2\tan z \tan w, \end{aligned}$$

or

$$\begin{aligned} &2 + \sec x \sec w + \sec y \sec z \\ &\leq \tan x \tan w + \tan y \tan z + (\tan x + \tan w)(\tan y + \tan z). \end{aligned}$$

Note that by the **addition and subtraction formulas,**

$$1 - \tan x \tan w = \frac{\cos x \cos w - \sin x \sin w}{\cos x \cos w} = \frac{\cos(x + w)}{\cos x \cos w}.$$

Hence,

$$1 - \tan x \tan w + \sec x \sec w = \frac{1 + \cos(x + w)}{\cos x \cos w}.$$

Similarly,

$$1 - \tan y \tan z + \sec y \sec z = \frac{1 + \cos(y + z)}{\cos y \cos z}.$$

Adding the last two equations gives

$$\begin{aligned} &2 + \sec x \sec w + \sec y \sec z - \tan x \tan w - \tan y \tan z \\ &= \frac{1 + \cos(x + w)}{\cos x \cos w} + \frac{1 + \cos(y + z)}{\cos y \cos z}. \end{aligned}$$

It suffices to show that

$$\frac{1 + \cos(x + w)}{\cos x \cos w} + \frac{1 + \cos(y + z)}{\cos y \cos z} \leq (\tan x + \tan w)(\tan y + \tan z),$$

or

$$s + t \leq (\tan x + \tan w)(\tan y + \tan z),$$

after setting $s = \frac{1+\cos(x+w)}{\cos x \cos w}$ and $t = \frac{1+\cos(y+z)}{\cos x \cos w}$. By the addition and subtraction formulas, we have

$$\tan x + \tan w = \frac{\sin x \cos w + \cos x \sin w}{\cos x \cos w} = \frac{\sin(x + w)}{\cos x \cos w}.$$

Similarly,

$$\tan y + \tan z = \frac{\sin(y+z)}{\cos y \cos z} = \frac{\sin(x+w)}{\cos y \cos z},$$

because $x + w = 180^\circ - (y + z)$. It follows that

$$\begin{aligned}
&(\tan x + \tan w)(\tan y + \tan z)\\
&\quad = \frac{\sin^2(x+w)}{\cos x \cos y \cos z \cos w} = \frac{1-\cos^2(x+w)}{\cos x \cos y \cos z \cos w}\\
&\quad = \frac{[1-\cos(x+w)][1+\cos(x+w)]}{\cos x \cos y \cos z \cos w}\\
&\quad = \frac{[1+\cos(y+z)][1+\cos(x+w)]}{\cos x \cos y \cos z \cos w} = st.
\end{aligned}$$

The desired inequality becomes $s+t \le st$, or $(1-s)(1-t) = 1-s-t+st \ge 1$. It suffices to show that $1 - s \ge 1$ and $1 - t \ge 1$. By symmetry, we have only to show that $1 - s \ge 1$; that is,

$$\frac{1+\cos(x+w)}{\cos x \cos w} \ge 2.$$

Multiplying both sides of the inequality by $\cos x \cos w$ and applying the addition and subtraction formulas gives

$$1 + \cos x \cos w - \sin x \sin w \ge 2 \cos x \cos w,$$

or $1 \ge \cos x \cos w + \sin x \sin w = \cos(x - w)$, which is evident. Equality holds if and only if $x = w$. Therefore, inequality $(*)$ is true, with equality if and only if $x = w$ and $y = z$, which happens precisely when $AD \parallel BC$ and $|AB| = |CD|$, as was to be shown.

Second Solution: We maintain the same notation as in the first solution. Applying the **law of cosines** to triangles ADI and BCI gives

$$|AI|^2 + |DI|^2 = 2\cos(x+w)|AI| \cdot |DI| + |AD|^2$$

and

$$|BI|^2 + |CI|^2 = 2\cos(y+z)|BI| \cdot |CI| + |BC|^2.$$

Adding the last two equations and completing squares gives

$$\begin{aligned}
&(|AI| + |DI|)^2 + (|BI| + |CI|)^2 + 2|AD| \cdot |BC|\\
&\quad = 2\cos(x+w)|AI| \cdot |DI| + 2\cos(y+z)|BI| \cdot |CI|\\
&\qquad + 2|AI| \cdot |DI| + 2|BI| \cdot |CI| + (|AD| + |BC|)^2.
\end{aligned}$$

Hence, establishing inequality (∗) is equivalent to establishing the inequality

$$[1+\cos(x+w)]|AI|\cdot|DI|+[1+\cos(y+z)]|BI|\cdot|CI|\le|AD|\cdot|BC|.$$

Since $2[ADI]=|AD|\cdot|ID_1|=|AI|\cdot|DI|\sin(x+w)$, $|AI|\cdot|DI|=\frac{|AD|}{\sin(x+w)}$. Similarly, $|BI|\cdot|CI|=\frac{|BC|}{\sin(y+z)}$. Because $x+w=180^\circ-(y+z)$, we have $\sin(x+w)=\sin(y+z)$ and $\cos(x+w)=-\cos(x+w)$. Plugging all the above information back into the last inequality yields

$$\frac{1+\cos(x+w)}{\sin(x+w)}\cdot|AD|+\frac{1-\cos(x+w)}{\sin(x+w)}\cdot|BC|\le|AD|\cdot|BC|,$$

or

$$\frac{1+\cos(x+w)}{|BC|}+\frac{1-\cos(x+w)}{|AD|}\le\sin(x+w). \qquad (**)$$

Note that by the addition and subtraction, the **product-to-sum**, and the **double-angle formulas**, we have

$$\begin{aligned}|AD|&=|AD_1|+|D_1D|=\tan x+\tan w=\frac{\sin x}{\cos x}+\frac{\sin w}{\cos w}\\&=\frac{\sin x\cos w+\cos x\sin w}{\cos x\cos w}=\frac{\sin(x+w)}{\cos x\cos w}=\frac{2\sin(x+w)}{2\cos x\cos w}\\&=\frac{4\sin\frac{x+w}{2}\cos\frac{x+w}{2}}{\cos(x+w)+\cos(x-w)}\ge\frac{4\sin\frac{x+w}{2}\cos\frac{x+w}{2}}{\cos(x+w)+1}\\&=\frac{4\sin\frac{x+w}{2}\cos\frac{x+w}{2}}{2\cos^2\frac{x+w}{2}}=2\tan\frac{x+w}{2}.\end{aligned}$$

Equality holds if and only if $\cos(x-w)=1$, that is, if $x=w$. (This step can be done easily by applying **Jensen's inequality**, using the fact $y=\tan x$ is convex for $0^\circ<x<90^\circ$.) Consequently, by the double-angle formulas,

$$\begin{aligned}\frac{1-\cos(x+w)}{|AD|}&\le\frac{2\sin^2\frac{x+w}{2}}{2\tan\frac{x+w}{2}}=\sin\frac{x+w}{2}\cos\frac{x+w}{2}\\&=\frac{\sin(x+w)}{2}.\end{aligned}$$

In exactly the same way, we can show that

$$\frac{1+\cos(x+w)}{|BC|}=\frac{1-\sin(y+z)}{|BC|}\le\frac{\sin(y+z)}{2}=\frac{\sin(x+w)}{2}.$$

Adding the last two inequalities gives the desired inequality (∗∗). Equality holds if and only if $x=w$ and $y=z$, which happens precisely when $AD\parallel BC$ and $|AB|=|CD|$, as was to be shown.

Third Solution: Because the circle ω is inscribed in $ABCD$, as shown in Figure 5.12, we can set $\angle DAI = \angle IAB = a$, $\angle ABI = \angle IBC = b$, $\angle BCI = \angle ICD = c$, $\angle CDI = \angle IDA = d$, and $a + b + c + d = 180°$. Our proof is based on the following key Lemma.

Lemma: *If a circle ω, centered at I, is inscribed in a quadrilateral $ABCD$, then*

$$|BI|^2 + \frac{|AI|}{|DI|} \cdot |BI| \cdot |CI| = |AB| \cdot |BC|. \qquad (\dagger)$$

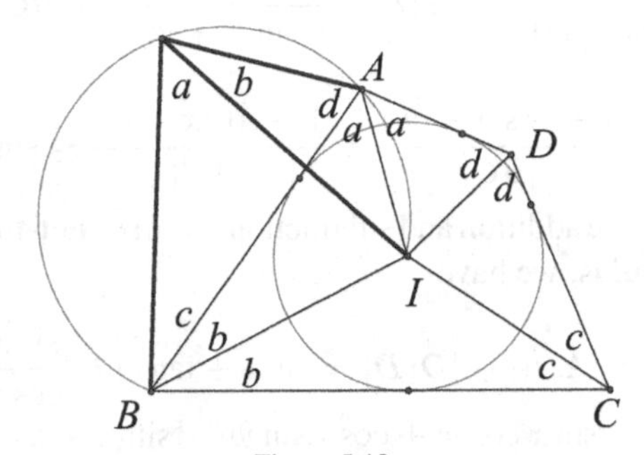

Figure 5.12.

Proof: Construct a point P outside of the quadrilateral such that triangle ABP is similar to triangle DCI. We obtain

$$\begin{aligned}\angle PAI + \angle PBI &= \angle PAB + \angle BAI + \angle PBA + \angle ABI \\ &= \angle IDC + a + \angle ICD + b \\ &= a + b + c + d = 180°,\end{aligned}$$

implying that the quadrilateral $PAIB$ is cyclic. By Ptolemy's theorem, we have $|AI| \cdot |BP| + |BI| \cdot |AP| = |AB| \cdot |IP|$, or

$$|BP| \cdot \frac{|AI|}{|IP|} + |BI| \cdot \frac{|AP|}{|IP|} = |AB|. \qquad (\dagger\dagger)$$

Because $PAIB$ is cyclic, it is not difficult to see that, as indicated in the figure, $\angle IPB = \angle IAB = a$, $\angle API = \angle ABI = b$, $\angle AIP = \angle ABP = c$, and $\angle PIB = \angle PAB = d$. Note that triangles AIP and ICB are similar, implying that

$$\frac{|AI|}{|IP|} = \frac{|IC|}{|CB|} \quad \text{and} \quad \frac{|AP|}{|IP|} = \frac{|IB|}{|CB|}.$$

Substituting the above equalities into the identity ($\dagger\dagger$), we arrive at

$$|BP| \cdot \frac{|CI|}{|BC|} + \frac{|BI|^2}{|BC|} = |AB|,$$

or

$$|BP| \cdot |CI| + |BI|^2 = |AB| \cdot |BC|. \qquad (\dagger\dagger\dagger)$$

Note also that triangle BIP and triangle IDA are similar, implying that $\frac{|BP|}{|BI|} = \frac{|IA|}{|ID|}$, or

$$|BP| = \frac{|AI|}{|ID|} \cdot |IB|.$$

Substituting the above identity back into ($\dagger\dagger\dagger$) gives the desired relation ($\dagger$), establishing the Lemma. ■

Now we prove our main result. By the Lemma and symmetry, we have

$$|CI|^2 + \frac{|DI|}{|AI|} \cdot |BI| \cdot |CI| = |CD| \cdot |BC|. \qquad (\ddagger)$$

Adding the two identities ($\dagger$) and ($\ddagger$) gives

$$|BI|^2 + |CI|^2 + \left(\frac{|AI|}{|DI|} + \frac{|DI|}{|AI|}\right)|BI| \cdot |CI| = |BC|(|AB| + |CD|).$$

By the **arithmetic–geometric means inequality**, we have $\frac{|AI|}{|DI|} + \frac{|DI|}{|AI|} \geq 2$. Thus,

$$|BC|(|AB| + |CD|) \geq |IB|^2 + |IC|^2 + 2|IB| \cdot |IC| = (|BI| + |CI|)^2,$$

where equality holds if and only if $|AI| = |DI|$. Likewise, we have

$$|AD|(|AB| + |CD|) \geq (|AI| + |DI|)^2,$$

where equality holds if and only if $|BI| = |CI|$. Adding the last two identities gives the desired inequality $(*)$.

By the given condition in the problem, all the equalities in the above discussion must hold; that is, $|AI| = |DI|$ and $|BI| = |CI|$. Consequently, we have $a = d$, $b = c$, and so $\angle DAB + \angle ABC = 2a + 2b = 180°$, implying that $AD \parallel BC$. It is not difficult to see that triangle AIB and triangle DIC are congruent, implying that $|AB| = |CD|$. Thus, $ABCD$ is an isosceles trapezoid.

44. [USAMO 2001] Let a, b, and c be nonnegative real numbers such that

$$a^2 + b^2 + c^2 + abc = 4.$$

Prove that

$$0 \leq ab + bc + ca - abc \leq 2.$$

The proof of the lower bound is rather simple. From the given condition, at least one of a, b, and c does not exceed 1, say $a \leq 1$. Then

$$ab + bc + ca - abc = a(b + c) + bc(1 - a) \geq 0.$$

To obtain equality, we have $a(b+c) = bc(1-a) = 0$. If $a = 1$, then $b+c = 0$ or $b = c = 0$, which contradicts the fact that $a^2 + b^2 + c^2 + abc = 4$. Hence $1 - a \neq 0$, and only one of b and c is 0. Without loss of generality, say $b = 0$. Therefore $b + c > 0$ and $a = 0$. Plugging $a = b = 0$ back into the given condition, we get $c = 2$. By permutation, the lower bound holds if and only if (a, b, c) is one of the triples $(2, 0, 0)$, $(0, 2, 0)$, and $(0, 0, 2)$. We next present three proofs of the upper bound.

First Solution: Based on Introductory Problem 22, we set $a = 2\sin\frac{A}{2}$, $b = 2\sin\frac{B}{2}$, and $c = 2\sin\frac{C}{2}$, where ABC is a triangle. We have

$$\begin{aligned} ab = 4\sin\frac{A}{2}\sin\frac{B}{2} &= 2\sqrt{\sin A\tan\frac{A}{2}\sin B\tan\frac{B}{2}} \\ &= 2\sqrt{\sin A\tan\frac{B}{2}\sin B\tan\frac{A}{2}}. \end{aligned}$$

By the **arithmetic–geometric means inequality**, this is at most

$$\begin{aligned} &\sin A\tan\frac{B}{2} + \sin B\tan\frac{A}{2} \\ &\quad = \sin A\cot\frac{A+C}{2} + \sin B\cot\frac{B+C}{2}. \end{aligned}$$

Likewise,

$$\begin{aligned} bc &\leq \sin B\cot\frac{B+A}{2} + \sin C\cot\frac{C+A}{2}, \\ ca &\leq \sin C\cot\frac{C+B}{2} + \sin A\cot\frac{A+B}{2}. \end{aligned}$$

Therefore, by the **sum-to-product**, **product-to-sum**, and the **double-angle formulas**, we have

$$
\begin{aligned}
&ab + bc + ca \\
&\quad \le (\sin A + \sin B) \cot \frac{A+B}{2} + (\sin B + \sin C) \cot \frac{B+C}{2} \\
&\quad + (\sin C + \sin A) \cot \frac{C+A}{2} \\
&\quad = 2 \cos \frac{A-B}{2} \cos \frac{A+B}{2} + 2 \cos \frac{B-C}{2} \cos \frac{B+C}{2} \\
&\quad + 2 \cos \frac{C-A}{2} \cos \frac{C+A}{2} \\
&\quad = 2(\cos A + \cos B + \cos C) \\
&\quad = 6 - 4 \left(\sin^2 \frac{A}{2} + \sin^2 \frac{B}{2} + \sin^2 \frac{C}{2} \right) \\
&\quad = 6 - \left(a^2 + b^2 + c^2 \right).
\end{aligned}
$$

Using the given equality, this last quantity equals $2 + abc$. It follows that

$$ab + bc + ca \le 2 + abc,$$

as desired.

Second Solution: Clearly, $0 \le a, b, c \le 2$. In the light of Introductory Problem 24(d), we can set $a = 2 \cos A$, $b = 2 \cos B$, and $c = 2 \cos C$, where ABC is an acute triangle. Either two of A, B, and C are at least 60°, or two of A, B, and C are at most 60°. Without loss of generality, assume that A and B have this property.

With these trigonometric substitutions, we find that the desired inequality is equivalent to

$$2(\cos A \cos B + \cos B \cos C + \cos C \cos A) \le 1 + 4 \cos A \cos B \cos C,$$

or,

$$
\begin{aligned}
&2(\cos A \cos B + \cos B \cos C + \cos C \cos A) \\
&\quad \le 3 - 2(\cos^2 A + \cos^2 B + \cos^2 C).
\end{aligned}
$$

Hence, by the **double-angle formulas**, it suffices to prove that

$$
\begin{aligned}
&\cos 2A + \cos 2B + \cos 2C \\
&+ 2(\cos A \cos B + 2 \cos B \cos C + \cos C \cos A) \le 0.
\end{aligned}
$$

By the sum-to-product and double-angle formulas, the sum of the first three terms in this inequality is

$$\begin{aligned}
&\cos 2A + \cos 2B + \cos 2C \\
&\qquad = 2\cos(A+B)\cos(A-B) + 2\cos^2(A+B) - 1 \\
&\qquad = 2\cos(A+B)[\cos(A-B) + \cos(A+B)] - 1 \\
&\qquad = 4\cos(A+B)\cos A\cos B - 1,
\end{aligned}$$

while the remaining terms equal

$$\begin{aligned}
&2\cos A\cos B + 2\cos C(\cos A + \cos B) \\
&\qquad = \cos(A+B) + \cos(A-B) - 2\cos(A+B)(\cos A + \cos B),
\end{aligned}$$

by the product-to-sum formulas. Hence, it suffices to prove that

$$\cos(A+B)[4\cos A\cos B + 1 - 2\cos A - 2\cos B] + \cos(A-B) \le 1,$$

or

$$-\cos C(1 - 2\cos A)(1 - 2\cos B) + \cos(A-B) \le 1. \qquad (*)$$

We consider the following cases:

(i) *At least one angle is* 60°. If A or B equals 60°, then we may assume, without loss of generality, that $A = 60^\circ$. If $C = 60^\circ$, then because either $A, B \ge 60^\circ$ or $A, B \le 60^\circ$, we must actually have $A = B = 60^\circ$, in which case equality holds. In either case, we may assume $A = 60^\circ$. Then $(*)$ becomes $\cos(A-B) \le 1$, which is always true, and equality holds if and only if $A = B = C = 60^\circ$, that is, if and only if $a = b = c = 1$.

(ii) *No angle equals* 60°. Because either $A, B \ge 60^\circ$ or $A, B \le 60^\circ$, we have $(1 - 2\cos A)(1 - 2\cos B) > 0$. Since $\cos C \ge 0$ and $\cos(A-B) \le 1$, $(*)$ is always true. Equality holds when $\cos C = 0$ and $\cos(A-B) = 1$. This holds exactly when $A = B = 45^\circ$ and $C = 90^\circ$; that is, when $a = b = \sqrt{2}$ and $c = 0$.

Third Solution: The problem also admits the following clever purely algebraic method, which is due to Oaz Nir and Richard Stong, independently.

Either two of a, b, c are less than or equal to 1, or two are greater than or equal to 1. Assume that b and c have this property. Then

$$b + c - bc = 1 - (1-b)(1-c) \le 1. \qquad (\dagger)$$

Viewing the given equality as a quadratic equation in a and solving for a yields

$$a = \frac{-bc \pm \sqrt{b^2c^2 - 4\left(b^2 + c^2\right) + 16}}{2}.$$

Note that

$$b^2c^2 - 4(b^2 + c^2) + 16 \le b^2c^2 - 8bc + 16 = (4 - bc)^2.$$

For the given equality to hold, we must have $b, c \le 2$, so that $4 - bc \ge 0$. Hence,

$$a \le \frac{-bc + |4 - bc|}{2} = \frac{-bc + 4 - bc}{2} = 2 - bc,$$

or

$$2 - bc \ge a. \qquad (\ddagger)$$

Combining the inequalities (†) and (‡) gives

$$2 - bc = (2 - bc) \cdot 1 \ge a(b + c - bc) = ab + ac - abc,$$

or $ab + ac + bc - abc \le 2$, as desired.

45. [Gabriel Dospinescu and Dung Tran Nam] Let s, t, u, v be numbers in the interval $\left(0, \frac{\pi}{2}\right)$ with $s + t + u + v = \pi$. Prove that

$$\frac{\sqrt{2}\sin s - 1}{\cos s} + \frac{\sqrt{2}\sin t - 1}{\cos t} + \frac{\sqrt{2}\sin u - 1}{\cos u} + \frac{\sqrt{2}\sin v - 1}{\cos v} \ge 0.$$

Solution: Set $a = \tan s$, $b = \tan t$, $c = \tan u$, and $d = \tan v$. Then a, b, c, d are positive real numbers. Because $s + t + u + v = \pi$, it follows that $\tan(s + t) + \tan(u + v) = 0$; that is,

$$\frac{a + b}{1 - ab} + \frac{c + d}{1 - cd} = 0,$$

by the **addition and subtraction formulas**. Multiplying both sides of the last equation by $(1 - ab)(1 - cd)$ yields

$$(a + b)(1 - cd) + (c + d)(1 - ab) = 0,$$

or

$$a + b + c + d = abc + bcd + cda + dab.$$

Consequently, we obtain

$$\begin{aligned}(a+b)(a+c)(a+d) &= a^2(a+b+c+d)+abc+bcd+cda+dab\\ &= (a^2+1)(a+b+c+d),\end{aligned}$$

or

$$\frac{a^2+1}{a+b}=\frac{(a+c)(a+d)}{a+b+c+d}$$

and its analogous forms. Hence

$$\begin{aligned}&\frac{a^2+1}{a+b}+\frac{b^2+1}{b+c}+\frac{c^2+1}{c+d}+\frac{d^2+1}{d+a}\\ &\quad=\frac{(a+c)(a+d)+(b+d)(b+a)+(c+a)(c+b)+(d+b)(d+c)}{a+b+c+d}\\ &\quad=\frac{a^2+b^2+c^2+d^2+2(ab+ac+ad+bc+bd+cd)}{a+b+c+d}\\ &\quad=a+b+c+d.\end{aligned}$$

By **Cauchy–Schwarz inequality**, we have

$$\begin{aligned}&2(a+b+c+d)^2\\ &\quad=2(a+b+c+d)\left(\frac{a^2+1}{a+b}+\frac{b^2+1}{b+c}+\frac{c^2+1}{c+d}+\frac{d^2+1}{d+a}\right)\\ &\quad=[(a+b)+(b+c)+(c+d)+(d+a)]\\ &\quad\quad\times\left(\frac{a^2+1}{a+b}+\frac{b^2+1}{b+c}+\frac{c^2+1}{c+d}+\frac{d^2+1}{d+a}\right)\\ &\quad\geq\left(\sqrt{a^2+1}+\sqrt{b^2+1}+\sqrt{c^2+1}+\sqrt{d^2+1}\right)^2,\end{aligned}$$

or

$$\sqrt{a^2+1}+\sqrt{b^2+1}+\sqrt{c^2+1}+\sqrt{d^2+1}\leq\sqrt{2}(a+b+c+d).$$

The least inequality is equivalent to

$$\frac{1}{\cos s}+\frac{1}{\cos t}+\frac{1}{\cos u}+\frac{1}{\cos v}\leq\sqrt{2}\left(\frac{\sin s}{\cos s}+\frac{\sin t}{\cos t}+\frac{\sin u}{\cos u}+\frac{\sin v}{\cos v}\right),$$

from which the desired inequality follows.

46. [USAMO 1995] Suppose a calculator is broken and the only keys that still work are the sin, cos, tan, $\sin^{-1}$, $\cos^{-1}$, and $\tan^{-1}$ buttons. The display initially shows 0. Given any positive rational number q, show that we can get q to appear on the display panel of the calculator by pressing some finite sequence of buttons. Assume that the calculator does real-number calculations with infinite precision, and that all functions are in terms of radians.

Solution: Because $\cos^{-1}\sin\theta = \frac{\pi}{2} - \theta$ and $\tan\left(\frac{\pi}{2} - \theta\right) = \frac{1}{\tan\theta}$ for $0 < \theta < \frac{\pi}{2}$, we have for any $x > 0$,

$$\tan\cos^{-1}\sin\tan^{-1}x = \tan\left(\frac{\pi}{2} - \tan^{-1}x\right) = \frac{1}{x}. \qquad (*)$$

Also, for $x \geq 0$,

$$\cos\tan^{-1}\sqrt{x} = \frac{1}{\sqrt{x+1}},$$

so by $(*)$,

$$\tan\cos^{-1}\sin\tan^{-1}\cos\tan^{-1}\sqrt{x} = \sqrt{x+1}. \qquad (**)$$

By induction on the denominator of r, we now prove that $\sqrt{r}$, for every nonnegative rational number r, can be obtained by using the operations

$$\sqrt{x} \mapsto \sqrt{x+1} \quad \text{and} \quad x \mapsto \frac{1}{x}.$$

If the denominator is 1, we can obtain $\sqrt{r}$, for every nonnegative integer r, by repeated application of $\sqrt{x} \mapsto \sqrt{x+1}$. Now assume that we can get $\sqrt{r}$ for all rational numbers r with denominator up to n. In particular, we can get any of

$$\sqrt{\frac{n+1}{1}},\ \sqrt{\frac{n+1}{2}},\ \ldots,\ \sqrt{\frac{n+1}{n}},$$

so we can also get

$$\sqrt{\frac{1}{n+1}},\ \sqrt{\frac{2}{n+1}},\ \ldots,\ \sqrt{\frac{n}{n+1}},$$

and $\sqrt{r}$, for any positive r of exact denominator $n+1$, can be obtained by repeatedly applying $\sqrt{x} \mapsto \sqrt{x+1}$.

Thus for any positive rational number r, we can obtain $\sqrt{r}$. In particular, we can obtain $\sqrt{q^2} = q$.

47. [China 2003, by Yumin Huang] Let n be a fixed positive integer. Determine the smallest positive real number λ such that for any $\theta_1, \theta_2, \ldots, \theta_n$ in the interval $(0, \frac{\pi}{2})$, if

$$\tan\theta_1 \tan\theta_2 \cdots \tan\theta_n = 2^{n/2},$$

then

$$\cos\theta_1 + \cos\theta_2 + \cdots + \cos\theta_n \le \lambda.$$

Solution: The answer is

$$\lambda = \begin{cases} \frac{\sqrt{3}}{3}, & n = 1; \\ \frac{2\sqrt{3}}{3}, & n = 2; \\ n - 1, & n \ge 3. \end{cases}$$

The case $n = 1$ is trivial. If $n = 2$, we claim that

$$\cos\theta_1 + \cos\theta_2 \le \frac{2\sqrt{3}}{3},$$

with equality if and only if $\theta_1 = \theta_2 = \tan^{-1}\sqrt{2}$. It suffices to show that

$$\cos^2\theta_1 + \cos^2\theta_2 + 2\cos\theta_1\cos\theta_2 \le \frac{4}{3},$$

or

$$\frac{1}{1+\tan^2\theta_1} + \frac{1}{1+\tan^2\theta_1} + 2\sqrt{\frac{1}{(1+\tan^2\theta_1)(1+\tan^2\theta_2)}} \le \frac{4}{3}.$$

Because $\tan\theta_1 \tan\theta_2 = 2$,

$$\left(1+\tan^2\theta_1\right)\left(1+\tan^2\theta_2\right) = 5 + \tan^2\theta_1 + \tan^2\theta_2.$$

By setting $\tan^2\theta_1 + \tan\theta_2^2 = x$, the last inequality becomes

$$\frac{2+x}{5+x} + 2\sqrt{\frac{1}{5+x}} \le \frac{4}{3},$$

or

$$2\sqrt{\frac{1}{5+x}} \le \frac{14+x}{3(5+x)}.$$

Squaring both sides and clearing denominators, we get $36(5+x) \le 196 + 28x + x^2$, that is, $0 \le x^2 - 8x + 16 = (x-4)^2$. This establishes our claim.

Now assume that $n \geq 3$. We claim that $\lambda = n - 1$. Note that $\lambda \geq n - 1$; by setting $\theta_2 = \theta_3 = \cdots = \theta_n = \theta$ and letting $\theta \to 0$, we find that $\theta_1 \to \frac{\pi}{2}$, and so the left-hand side of the desired inequality approaches $n - 1$. It suffices to show that

$$\cos\theta_1 + \cos\theta_2 + \cdots + \cos\theta_n \leq n - 1.$$

Without loss of generality, assume that $\theta_1 \geq \theta_2 \geq \cdots \geq \theta_n$. Then

$$\tan\theta_1 \tan\theta_2 \tan\theta_3 \geq 2\sqrt{2}.$$

It suffices to show that

$$\cos\theta_1 + \cos\theta_2 + \cos\theta_3 < 2. \qquad (*)$$

Because $\sqrt{1 - x^2} \leq 1 - \frac{1}{2}x^2$, $\cos\theta_i = \sqrt{1 - \sin^2\theta_i} < 1 - \frac{1}{2}\sin^2\theta_i$. Consequently, by the **arithmetic–geometric means inequality**,

$$\cos\theta_2 + \cos\theta_3 < 2 - \frac{1}{2}\left(\sin^2\theta_2 + \sin^2\theta_3\right) \leq 2 - \sin\theta_2 \sin\theta_3.$$

Because

$$\tan^2\theta_1 \geq \frac{8}{\tan^2\theta_2 \tan^2\theta_3},$$

we have

$$\sec^2\theta_1 \geq \frac{8 + \tan^2\theta_2 \tan^2\theta_3}{\tan^2\theta_2 \tan^2\theta_3},$$

or

$$\cos\theta_1 \leq \frac{\tan\theta_2 \tan\theta_3}{\sqrt{8 + \tan^2\theta_2 \tan^2\theta_3}} = \frac{\sin\theta_2 \sin\theta_3}{\sqrt{8\cos^2\theta_2 \cos^2\theta_3 + \sin^2\theta_2 \sin^2\theta_3}}.$$

It follows that

$$\begin{aligned}&\cos\theta_1 + \cos\theta_2 + \cos\theta_3 \\ &\quad < 2 - \sin\theta_2 \sin\theta_3 \left[1 - \frac{1}{8\cos^2\theta_2 \cos^2\theta_3 + \sin^2\theta_2 \sin^2\theta_3}\right].\end{aligned}$$

It is clear that the equality

$$8\cos^2\theta_2 \cos^2\theta_3 + \sin^2\theta_2 \sin^2\theta_3 \geq 1, \qquad (**)$$

establishes the desired inequality $(*)$. Inequality $(**)$ is equivalent to

$$8 + \tan^2\theta_2 \tan^2\theta_3 \geq (1 + \tan^2\theta_2)(1 + \tan^2\theta_3),$$

or

$$7 \ge \tan^2\theta_2 + \tan^2\theta_3.$$

Thus if $\tan^2\theta_2 + \tan^2\theta_3 \le 7$, then inequality $(*)$ holds, and we are done. Assume that $\tan^2\theta_2 + \tan^2\theta_3 > 7$. Then $\tan^2\theta_1 \ge \tan^2\theta_2 \ge \frac{7}{2}$. Then

$$\cos\theta_1 \le \cos\theta_2 = \frac{1}{\sqrt{1+\tan^2\theta_2}} \le \frac{\sqrt{2}}{3},$$

implying that

$$\cos\theta_1 + \cos\theta_2 + \cos\theta_3 \le \frac{2\sqrt{2}}{3} + 1 < 2,$$

establishing $(*)$ again.

Therefore, inequality $(*)$ is true, as desired.

48. Let ABC be an acute triangle. Prove that

$$(\sin 2B + \sin 2C)^2 \sin A + (\sin 2C + \sin 2A)^2 \sin B$$
$$+ (\sin 2A + \sin 2B)^2 \sin C \le 12 \sin A \sin B \sin C.$$

First Solution: Applying the **addition and subtraction formulas** gives

$$\begin{aligned}(\sin 2B + \sin 2C)^2 \sin A &= 4\sin^2(B+C)\cos^2(B-C)\sin A \\ &= 4\sin^3 A\cos^2(B-C),\end{aligned}$$

because $A + B + C = 180°$. Hence it suffices to show that the **cyclic sum**

$$\sum_{\text{cyc}} \sin^3 A \cos^2(B-C)$$

is less than or equal to $3\sin A \sin B \sin C$, which follows from

$$\sum_{\text{cyc}} 4\sin^3 A\cos(B-C) = 12\sin A\sin B\sin C.$$

Indeed, we have

$$\begin{aligned}&4\sin^3 A\cos(B-C)\\ &= 4\sin^2 A\sin(B+C)\cos(B-C)\\ &= 2\sin^2 A(\sin 2B + \sin 2C)\\ &= (1-\cos 2A)(\sin 2B + \sin 2C)\\ &= (\sin 2B + \sin 2C) - \sin 2B\cos 2A - \sin 2C\cos 2A.\end{aligned}$$

It follows that

$$
\begin{aligned}
&\sum_{\text{cyc}} 4\sin^3 A\cos(B-C)\\
&\quad=\sum_{\text{cyc}}(\sin 2B+\sin 2C)-\sum_{\text{cyc}}\sin 2B\cos 2A-\sum_{\text{cyc}}\sin 2C\cos 2A\\
&\quad=2\sum_{\text{cyc}}\sin 2A-\sum_{\text{cyc}}\sin 2B\cos 2A-\sum_{\text{cyc}}\sin 2A\cos 2B\\
&\quad=2\sum_{\text{cyc}}\sin 2A-\sum_{\text{cyc}}(\sin 2B\cos 2A+\sin 2A\cos 2B)\\
&\quad=2\sum_{\text{cyc}}\sin 2A-\sum_{\text{cyc}}\sin(2B+2A)\\
&\quad=2\sum_{\text{cyc}}\sin 2A+\sum_{\text{cyc}}\sin 2C\\
&\quad=3(\sin 2A+\sin 2B+\sin 2C)\\
&\quad=12\sin A\sin B\sin C,
\end{aligned}
$$

by Introductory Problem 24(a). Equality holds if and only if $\cos(A-B)=\cos(B-C)=\cos(C-A)=1$, that is, if and only if triangle ABC is equilateral.

Note: Enlarging $\sin^3 A\cos^2(B-C)$ to $\sin^3 A\cos(B-C)$ is a very clever but somewhat tricky idea. The following more geometric approach reveals more of the motivation behind the problem. Please note the last step in the proof of the Lemma below.

Second Solution: We can rewrite the desired inequality as

$$\sum_{\text{cyc}}(\sin 2B+\sin 2C)^2\sin A\le 12\sin A\sin B\sin C.$$

By the **extended law of sines**, we have $c=2R\sin C$, $a=2R\sin A$, and $b=2R\sin B$. Hence

$$12R^2\sin A\sin B\sin C=3ab\sin C=6[ABC].$$

It suffices to show that

$$R^2\sum_{\text{cyc}}(\sin 2B+\sin 2C)^2\sin A\le 6[ABC].\qquad (*)$$

We establish the following Lemma.

Lemma *Let AD, BE, CF be the altitudes of acute triangle ABC, with D, E, F on sides BC, CA, AB, respectively. Then*

$$|DE| + |DF| \le |BC|.$$

Equality holds if and only if $|AB| = |AC|$.

Proof: We consider Figure 5.13. Because $\angle CFA = \angle CDA = 90^\circ$, quadrilateral $AFDC$ is cyclic, and so $\angle FDB = \angle BAC = \angle CAB$ and $\angle BFD = \angle BCA = \angle BCA$. Hence triangles BDF and BAC are similar, so

$$\frac{|DF|}{|AC|} = \frac{|BF|}{|BC|} = \cos B,$$

or (by the **double-angle formula**)

$$|DF| = b\cos B = 2R\sin B\cos B = R\sin 2B.$$

Likewise, $|DE| = c\cos C = R\sin 2C$. Thus,

$$|DE| + |DF| = R(\sin 2B + \sin 2C). \tag{†}$$

Since $0^\circ < A, B, C < 90^\circ$, by the **sum-to-product formula**,

$$\begin{aligned} |BC| - (|DE| + |DF|) &= R[2\sin A - (\sin 2B + \sin 2C] \\ &= R[2\sin A - 2\sin(B + C)\cos(B - C)] \\ &= 2R\sin A[1 - \cos(B - C)] \ge 0, \end{aligned}$$

as desired. (This can also be proven by the **law of cosines**.) ■

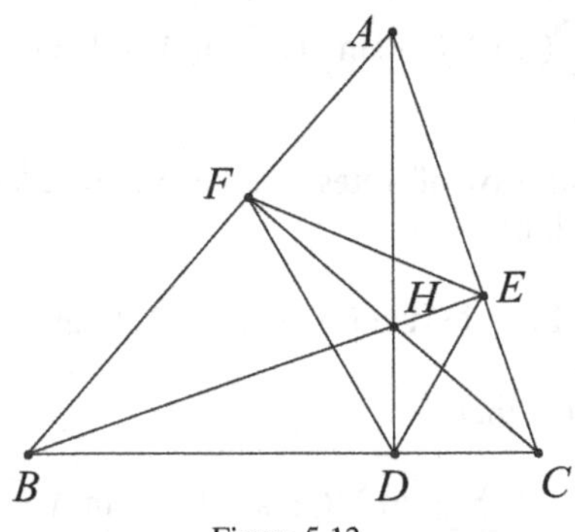

Figure 5.13.

Because both $ABDE$ and $ACDF$ are cyclic, $\angle BDF = \angle CDE = \angle CAB$. Thus, by the Lemma, we have

$$\begin{aligned}
&2([BFC] + [BEC]) \\
&\quad = |DF| \cdot |BC| \cdot \sin \angle BDF + |DE| \cdot |BC| \cdot \sin \angle EDC \\
&\quad = |BC|(|DE| + |DF|) \sin A \geq (|DE| + |DF|)^2 \sin A.
\end{aligned}$$

By equation (†), the last inequality is equivalent to

$$R^2(\sin 2B + \sin 2C)^2 \sin A \leq 2[BFC] + 2[BEC].$$

Likewise, we have

$$R^2(\sin 2C + \sin 2A)^2 \sin B \leq 2[CDA] + 2[CFA]$$

and

$$R^2(\sin 2A + \sin 2B)^2 \sin C \leq 2[AEB] + 2[ADB].$$

Adding the last three inequalities yields the desired result. In view of the Lemma, it is also clear that equality holds if and only if triangle ABC is equilateral.

49. [Bulgaria 1998] On the sides of a nonobtuse triangle ABC are constructed externally a square P_4, a regular m-sided polygon P_m, and a regular n-sided polygon P_n. The centers of the square and the two polygons form an equilateral triangle. Prove that $m = n = 6$, and find the angles of triangle ABC.

Solution: The angles are $90°$, $45°$, and $45°$. We prove the following lemma.

Lemma *Let O be a point inside equilateral triangle XYZ. If*

$$\angle YOZ = x,\ \angle ZOX = y,\ \angle XOY = z.$$

then

$$\frac{|OX|}{\sin(x - 60°)} = \frac{|OY|}{\sin(y - 60°)} = \frac{|OZ|}{\sin(z - 60°)}.$$

Proof: As shown in Figure 5.14, let $\mathbf{R}$ denote clockwise rotation of $60°$ around the point Z, and let $\mathbf{R}(X) = X_1$ and $\mathbf{R}(O) = O_1$.

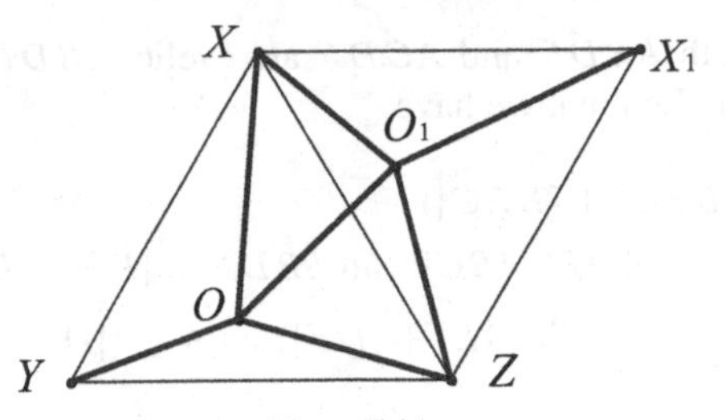

Figure 5.14.

Then $\mathbf{R}(Y) = X$, and so triangle ZO_1O is equilateral. Consequently, triangles ZO_1X and ZOY are congruent, and so $|O_1X| = |OY|$. Note that $x + y + z = 360°$. We have

$$\begin{aligned}
\angle O_1OX &= \angle ZOX - \angle ZOO_1 = \angle ZOX - 60° = y - 60°,\\
\angle XO_1O &= \angle XO_1Z - \angle OO_1Z = \angle YOZ - 60° = x - 60°,\\
\angle OXO_1 &= 180° - \angle O_1OX - \angle XO_1O = z - 60°.
\end{aligned}$$

Applying the **law of sines** to triangle XOO_1 establishes the desired result. ■

Now we prove our main result. Without loss of generality, suppose that P_4, P_m, and P_n are on sides AB, BC, and CA, respectively (Figure 5.15). Let O be the circumcenter of triangle ABC. Without loss of generality, we assume that the circumradius of triangle ABC is 1, so $|OA| = |OB| = |OC| = 1$. Let X, Y, and Z be the centers of P_4, P_m, and P_n, respectively.

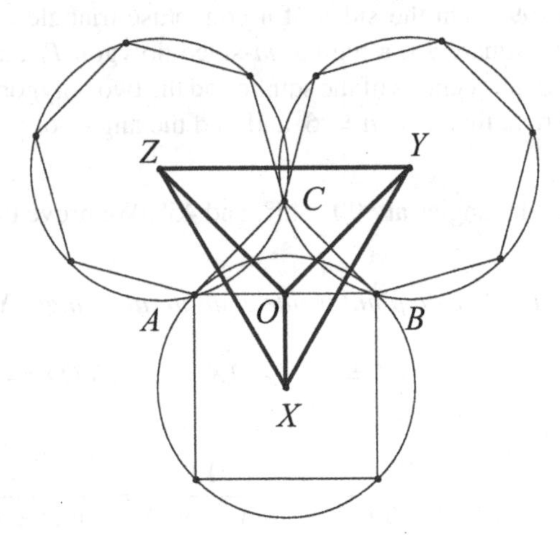

Figure 5.15.

Because $|OB| = |OC|$ and $|YB| = |YC|$, $BOCY$ is a **kite** with OY as its axis of symmetry. Thus, $\angle BOY = \frac{\angle BOC}{2} = \angle A$, and $\angle OYB = 180°/m$. Let

$\alpha = \frac{180^\circ}{m}$. Applying the law of sines to triangle OBY, we obtain

$$|OY| = \frac{\sin(A+\alpha)}{\sin\alpha}.$$

Likewise, by setting $\angle ZOC = \frac{180^\circ}{n} = \beta$, we have

$$|OX| = \frac{\sin(C+45^\circ)}{\sin 45^\circ} = \sqrt{2}\sin(C+45^\circ) \quad \text{and} \quad |OZ| = \frac{\sin(B+\beta)}{\sin\beta}.$$

Note that O is inside triangle XYZ, because it is on the respective perpendicular rays from X, Y, and Z toward sides AB, BC, and CA. Because $\angle BOY = \angle A$ and $\angle BOX = \angle C$, we find that $\angle XOY = \angle C + \angle A$. Likewise, $\angle YOZ = \angle A + \angle B$ and $\angle ZOX = \angle B + \angle C$. Applying the Lemma gives

$$\frac{|OY|}{\sin(B+C-60^\circ)} = \frac{|OZ|}{\sin(C+A-60^\circ)} = \frac{|OX|}{\sin(A+B-60^\circ)},$$

or

$$\frac{|OY|}{\sin(A+60^\circ)} = \frac{|OZ|}{\sin(B+60^\circ)} = \frac{|OX|}{\sin(C+60^\circ)}.$$

It follows that

$$\frac{\sin(A+\alpha)\csc\alpha}{\sin(A+60^\circ)} = \frac{\sin(B+\beta)\csc\beta}{\sin(B+60^\circ)} = \frac{\sqrt{2}\sin(C+45^\circ)}{\sin(C+60^\circ)}.$$

Because $y = \cot x$ is decreasing for x with $0^\circ \le x \le 180^\circ$, by the **addition and subtraction formulas**, the function

$$f(x) = \frac{\sin(x-15^\circ)}{\sin x} = \cos 15^\circ - \cot x \sin 15^\circ$$

is increasing for x with $0^\circ \le x \le 90^\circ$. Consequently, because $0^\circ \le C+60^\circ \le 150^\circ$ ($\angle C \le 90^\circ$), it follows that

$$\frac{\sqrt{2}\sin(C+45^\circ)}{\sin(C+60^\circ)} \le \frac{\sqrt{2}\sin(90^\circ+45^\circ)}{\sin(90^\circ+60^\circ)} = 2,$$

with equality if and only if $C = 90^\circ$. Therefore,

$$\frac{\sin(A+\alpha)\csc\alpha}{\sin(A+60^\circ)} = \frac{\sin(B+\beta)\csc\beta}{\sin(B+60^\circ)} \le 2. \qquad (*)$$

Because triangle ABC is nonobtuse, at least two of its angles are between 45° and 90°. Without loss of generality, we may assume that $45^\circ \le B$ ($\le 90^\circ$), so $\sin(B + 60^\circ) > 0$ and $\cot B \leq 1$. Then from the second inequality in the relation $(*)$, we have

$$\sin(B + \beta)\csc\beta \leq 2\sin(B + 60^\circ),$$

or

$$\sin B \cot\beta + \cos B \leq \sin B + \sqrt{3}\cos B,$$

by the addition and subtraction formulas. Dividing both sides of the above inequality by $\sin B$ yields

$$\cot\beta \leq 1 + (\sqrt{3} - 1)\cot B \leq 1 + \sqrt{3} - 1 = \sqrt{3},$$

implying that $\beta \geq 30^\circ$. But since $n \geq 6$, $\beta = 180^\circ/n \leq 30^\circ$. Thus all the equalities hold, and so $\angle C = 90^\circ$ and $\angle A = \angle B = 45^\circ$, as claimed.

50. [MOSP 2000] Let ABC be an acute triangle. Prove that

$$\left(\frac{\cos A}{\cos B}\right)^2 + \left(\frac{\cos B}{\cos C}\right)^2 + \left(\frac{\cos C}{\cos A}\right)^2 + 8\cos A\cos B\cos C \geq 4.$$

Note: It is easier to rewrite the above inequality in terms of $\cos^2 A$, $\cos^2 B$, and $\cos^2 C$. By Introductory Problem 24(d), we have

$$4 - 8\cos A\cos B\cos C = 4\left(\cos^2 A + \cos^2 B + \cos^2 C\right).$$

It suffices to prove

$$\left(\frac{\cos A}{\cos B}\right)^2 + \left(\frac{\cos B}{\cos C}\right)^2 + \left(\frac{\cos C}{\cos A}\right)^2 \geq 4\left(\cos^2 A + \cos^2 B + \cos^2 C\right). \quad (\dagger)$$

We present three approaches.

First Solution: By the **weighted arithmetic–geometric means inequality**, we have

$$\begin{aligned}
2\left(\frac{\cos A}{\cos B}\right)^2 + \left(\frac{\cos B}{\cos C}\right)^2 &\geq 3\sqrt[3]{\frac{\cos^4 A}{\cos^2 B\cos^2 C}} \\
&= \frac{3\cos^2 A}{\sqrt[3]{\cos^2 A\cos^2 B\cos^2 C}} \\
&\geq 12\cos^2 A,
\end{aligned}$$

by Introductory Problem 28(a). Adding the above inequality with its analogous forms and dividing both sides of the resulting inequality by 3, we obtain inequality (†).

Second Solution: Setting $x = \frac{\cos B}{\cos C}$, $y = \frac{\cos C}{\cos A}$, $z = \frac{\cos A}{\cos B}$ in Advanced Problem 42(a) yields

$$\begin{aligned}
&\left(\frac{\cos A}{\cos B}\right)^2 + \left(\frac{\cos B}{\cos C}\right)^2 + \left(\frac{\cos C}{\cos A}\right)^2 \\
&\quad = x^2 + y^2 + z^2 \\
&\quad \geq 2(yz \cos A + zx \cos B + xy \cos C) \\
&\quad = 2\left[\frac{\cos C \cos A}{\cos B} + \frac{\cos A \cos B}{\cos C} + \frac{\cos B \cos C}{\cos A}\right].
\end{aligned}$$

However, setting

$$x = \sqrt{\frac{\cos B \cos C}{\cos A}}, \quad y = \sqrt{\frac{\cos A \cos B}{\cos C}}, \quad z = \sqrt{\frac{\cos C \cos A}{\cos B}},$$

in Advanced Problem 42(a) again, we find that

$$\begin{aligned}
&2\left[\frac{\cos C \cos A}{\cos B} + \frac{\cos A \cos B}{\cos C} + \frac{\cos B \cos C}{\cos A}\right] \\
&\quad = 2(x^2 + y^2 + z^2) \\
&\quad \geq 4(yz \cos A + zx \cos B + xy \cos C) \\
&\quad = 4\left(\cos^2 A + \cos^2 B + \cos^2 C\right),
\end{aligned}$$

by noting that

$$yz \cos A = \cos A \sqrt{\frac{\cos A \cos B}{\cos C} \cdot \frac{\cos C \cos A}{\cos B}} = \cos^2 A$$

and its analogous forms for $zx \cos B$ and $xy \cos C$.

Third Solution: The result follows from the following Lemma.

Lemma *For positive real numbers a, b, c such that $abc \leq 1$,*

$$\frac{a}{b} + \frac{b}{c} + \frac{c}{a} \geq a + b + c.$$

Proof: Replacing a, b, c by ta, tb, tc with $t = 1/\sqrt[3]{abc}$ leaves the left-hand side of the inequality unchanged and increases the value of the right-hand side and results in the equality $atbtct = abct^3 = 1$. Hence we may assume without loss of generality that $abc = 1$. Then there exist positive real numbers x, y, z such that $a = x/y, b = z/x, c = y/z$. The **rearrangement inequality** gives

$$x^3 + y^3 + z^3 \geq x^2z + y^2x + z^2y.$$

Thus

$$\begin{aligned}\frac{a}{b} + \frac{b}{c} + \frac{c}{a} &= \frac{x^2}{yz} + \frac{y^2}{zx} + \frac{z^2}{xy} = \frac{x^3 + y^3 + z^3}{xyz} \\ &\geq \frac{x^2z + y^2x + z^2y}{xyz} = \frac{x}{y} + \frac{y}{z} + \frac{z}{x} \\ &= a + b + c,\end{aligned}$$

as desired. ■

Now we prove our main result. Note that

$$\left(4\cos^2 A\right)\left(4\cos^2 B\right)\left(4\cos^2 C\right) = (8\cos A\cos B\cos C)^2 \leq 1$$

by Introductory Problem 28(a). Setting $a = 4\cos^2 A$, $b = 4\cos^2 B$, $c = 4\cos^2 C$ in the Lemma yields

$$\begin{aligned}\left(\frac{\cos A}{\cos B}\right)^2 + \left(\frac{\cos B}{\cos C}\right)^2 + \left(\frac{\cos C}{\cos A}\right)^2 &= \frac{a}{b} + \frac{b}{c} + \frac{c}{a} \geq a + b + c \\ &= 4(\cos^2 A + \cos^2 B + \cos^2 C),\end{aligned}$$

establishing inequality (†).

51. For any real number x and any positive integer n, prove that

$$\left|\sum_{k=1}^{n} \frac{\sin kx}{k}\right| \leq 2\sqrt{\pi}.$$

Solution: The solution is based on the following three Lemmas.

Lemma 1 *Let n be a positive integer, and let $a_1, a_2, \ldots, a_n$ and $b_1, b_2, \ldots, b_n$ be two sequences of real numbers. Then*

$$\sum_{k=1}^{n} a_k b_k = S_n b_n + \sum_{k=1}^{n-1} S_k(b_k - b_{k+1}),$$

where $S_k = a_1 + a_2 + \cdots + a_k$, *for* $k = 1, 2, \ldots, n$.

Proof: Set $S_0 = 0$. Then $a_k = S_k - S_{k-1}$ for $k = 1, 2, \ldots, n$, and so

$$\begin{aligned}\sum_{k=1}^{n} a_k b_k &= \sum_{k=1}^{n} (S_k - S_{k-1}) b_k = \sum_{k=1}^{n} S_k b_k - \sum_{k=1}^{n} S_{k-1} b_k \\ &= S_n b_n + \sum_{k=1}^{n-1} S_k b_k - \sum_{k=2}^{n} S_{k-1} b_k - S_0 b_1 \\ &= S_n b_n + \sum_{k=1}^{n-1} S_k b_k - \sum_{k=1}^{n-1} S_k b_{k+1} \\ &= S_n b_n + \sum_{k=1}^{n-1} S_k (b_k - b_{k+1}),\end{aligned}$$

as desired. ■

Lemma 2 *[Abel's inequality] Let* n *be a positive integer, and let* $a_1, a_2, \ldots, a_n$ *and* $b_1, b_2, \ldots, b_n$ *be two sequences of real numbers with* $b_1 \ge b_2 \ge \cdots \ge b_n \ge 0$. *Then*

$$m b_1 \le \sum_{k=1}^{n} a_k b_k \le M b_1,$$

where $S_k = a_1 + a_2 + \cdots + a_k$, *for* $k = 1, 2, \ldots, n$, *and* M *and* m *are the maximum and minimum, respectively, of* $\{ S_1, S_2, \ldots, S_n \}$.

Proof: Note that $b_n \ge 0$ and $b_k - b_{k+1} \ge 0$ for $k = 1, 2, \ldots, n-1$. Lemma 1 gives

$$\begin{aligned}\sum_{k=1}^{n} a_k b_k &= S_n b_n + \sum_{k=1}^{n-1} S_k (b_k - b_{k+1}) \\ &\le M b_n + M \sum_{k=1}^{n-1} (b_k - b_{k+1}) = M b_1.\end{aligned}$$

establishing the second desired inequality. In exactly the same way, we can prove the first desired inequality. ■

Lemma 3 *Let* x *be a real number that is not an even multiple of* π, *then*

$$\left| \sum_{k=m+1}^{n} \frac{\sin kx}{k} \right| \le \frac{1}{(m+1)\left|\sin \frac{x}{2}\right|},$$

where m and n are positive integers with $m < n$.

Proof: For $k = 1, 2, \ldots, n-m$, let $a_k = \sin[(k+m)x]\sin\frac{x}{2}$ and $b_k = \frac{1}{k+m}$. Then by Lemma 2, we have

$$\frac{s}{m+1} = sb_1 \le \sum_{k=m+1}^{n} \frac{\sin kx \sin\frac{x}{2}}{k} = \sum_{k=1}^{n-m} a_k b_k \le Sb_1 = \frac{S}{m+1},$$

where $S_k = a_1 + a_2 + \cdots + a_k$, and $S = \max\{S_1, S_2, \ldots S_n\}$ and $s = \min\{S_1, S_2, \ldots, S_n\}$. The **product-to-sum formulas** give

$$\begin{aligned}2a_i &= 2\sin[(i+m)x]\sin\frac{x}{2}\\ &= \cos\left(i+m-\frac{1}{2}\right)x - \cos\left(i+m+\frac{1}{2}\right)x,\end{aligned}$$

and so

$$2S_k = 2a_1 + 2a_2 + \cdots + 2a_k = \cos\left(m+\frac{1}{2}\right)x - \cos\left(k+m+\frac{1}{2}\right)x.$$

It follows that $-2 \le 2S_k \le 2$ for $k = 1, 2, \ldots, n$, and so $-1 \le s \le S \le 1$. Consequently,

$$-\frac{1}{m+1} = -b_1 \le sb_1 \le \sum_{k=m+1}^{n} \frac{\sin kx \sin\frac{x}{2}}{k} \le Sb_1 \le b_1 = \frac{1}{m+1},$$

implying that

$$\left|\sum_{k=m+1}^{n} \frac{\sin kx \sin\frac{x}{2}}{k}\right| \le \frac{1}{m+1},$$

from which the desired result follows. ■

Now we are ready to prove our main result. Because $y = |\sin x|$ is a periodic function with period π, we may assume that x is in the interval $(0, \pi)$. (Note that the desired result is trivial for $x = 0$.) For a fixed x with $0 < x < \pi$, let m be the nonnegative integer such that

$$m \le \frac{\sqrt{\pi}}{x} < m + 1.$$

Thus

$$\left|\sum_{k=1}^{n} \frac{\sin kx}{k}\right| \le \left|\sum_{k=1}^{m} \frac{\sin kx}{k}\right| + \left|\sum_{k=m+1}^{n} \frac{\sin kx}{k}\right|.$$

Here we set the first summation on the right-hand side to be 0 if $m = 0$, and the first summation taken from 1 to n and the second to be 0 if $m \geq n$. It suffices to show that

$$\left|\sum_{k=1}^{m} \frac{\sin kx}{k}\right| \leq \sqrt{\pi} \tag{$*$}$$

and

$$\left|\sum_{k=m+1}^{n} \frac{\sin kx}{k}\right| \leq \sqrt{\pi}. \tag{$**$}$$

Because $|\sin x| < x$ and by the definition of m, it follows that

$$\left|\sum_{k=1}^{m} \frac{\sin kx}{k}\right| \leq \sum_{k=1}^{m} \frac{kx}{k} = \sum_{k=1}^{m} x = mx \leq \sqrt{\pi},$$

establishing inequality ($*$). On the other hand, by Lemma 3, we have

$$\left|\sum_{k=m+1}^{n} \frac{\sin kx}{k}\right| \leq \frac{1}{(m+1)\left|\sin \frac{x}{2}\right|}.$$

Note that $y = \sin x$ is concave for $0 < x < \frac{\pi}{2}$. Thus, the graph of $y = \sin x$ is above the line connecting the points $(0, 0)$ and $\left(\frac{\pi}{2}, 1\right)$ on the interval $\left(0, \frac{\pi}{2}\right)$; that is, $\sin x > \frac{2x}{\pi}$. Hence for $0 < x < \pi$, we have

$$\sin \frac{x}{2} < \frac{2 \cdot \frac{x}{2}}{\pi} = \frac{x}{\pi}.$$

It follows that

$$\left|\sum_{k=m+1}^{n} \frac{\sin kx}{k}\right| \leq \frac{1}{(m+1)\left|\sin \frac{x}{2}\right|} \leq \frac{1}{m+1} \cdot \frac{x}{\pi} \leq \frac{\sqrt{\pi}}{x} \cdot \frac{x}{\pi} = \sqrt{\pi},$$

establishing inequality ($**$). Our proof is thus complete.

Glossary

Arithmetic–Geometric Means Inequality

If n is a positive integer and $a_1, a_2, \ldots, a_n$ are nonnegative real numbers, then

$$\frac{1}{n}\sum_{i=1}^{n} a_i \geq (a_1 a_2 \cdots a_n)^{1/n},$$

with equality if and only if $a_1 = a_2 = \cdots = a_n$. This inequality is a special case of the **power mean inequality**.

Arithmetic–Harmonic Means Inequality

If $a_1, a_2, \ldots, a_n$ are n positive numbers, then

$$\frac{1}{n}\sum_{i=1}^{n} a_i \geq \frac{1}{\frac{1}{n}\sum_{i=1}^{n}\frac{1}{a_i}},$$

with equality if and only if $a_1 = a_2 = \cdots = a_n$. This inequality is a special case of the **power mean inequality**.

Binomial Coefficient

$$\binom{n}{k} = \frac{n!}{k!(n-k)!},$$

the coefficient of x^k in the expansion of $(x+1)^n$.

Cauchy–Schwarz Inequality

For any real numbers $a_1, a_2, \ldots, a_n$, and $b_1, b_2, \ldots, b_n$

$$(a_1^2 + a_2^2 + \cdots + a_n^2)(b_1^2 + b_2^2 + \cdots + b_n^2)$$
$$\geq (a_1b_1 + a_2b_2 + \cdots + a_nb_n)^2,$$

with equality if and only if a_i and b_i are proportional, $i = 1, 2, \ldots, n$.

Ceva's Theorem and Its Trigonometric Form

Let AD, BE, CF be three **cevians** of triangle ABC. The following are equivalent:

(i) AD, BE, CF are concurrent;

(ii) $\frac{|AF|}{|FB|} \cdot \frac{|BD|}{|DC|} \cdot \frac{|CE|}{|EA|} = 1;$

(iii) $\frac{\sin \angle ABE}{\sin \angle EBC} \cdot \frac{\sin \angle BCF}{\sin \angle FCA} \cdot \frac{\sin \angle CAD}{\sin \angle DAB} = 1.$

Cevian

A cevian of a triangle is any segment joining a vertex to a point on the opposite side.

Chebyshev's Inequality

1. Let $x_1, x_2, \ldots, x_n$ and $y_1, y_2, \ldots, y_n$ be two sequences of real numbers such that $x_1 \leq x_2 \leq \cdots \leq x_n$ and $y_1 \leq y_2 \leq \cdots \leq y_n$. Then

$$\frac{1}{n}(x_1 + x_2 + \cdots + x_n)(y_1 + y_2 + \cdots + y_n) \leq x_1y_1 + x_2y_2 + \cdots + x_ny_n.$$

2. Let $x_1, x_2, \ldots, x_n$ and $y_1, y_2, \ldots, y_n$ be two sequences of real numbers such that $x_1 \geq x_2 \geq \cdots \geq x_n$ and $y_1 \geq y_2 \geq \cdots \geq y_n$. Then

$$\frac{1}{n}(x_1 + x_2 + \cdots + x_n)(y_1 + y_2 + \cdots + y_n) \geq x_1y_1 + x_2y_2 + \cdots + x_ny_n.$$

Chebyshev Polynomials

Let $\{T_n(x)\}_{n=0}^{\infty}$ be the sequence of polynomials such that $T_0(x) = 1$, $T_1(x) = x$, and $T_{i+1} = 2xT_i(x) - T_{i-1}(x)$ for all positive integers i. The polynomial $T_n(x)$ is called the nth Chebyshev polynomial.

Circumcenter

The center of the circumscribed circle or sphere.

Circumcircle

A circumscribed circle.

Convexity

A function $f(x)$ is **concave up (down)** on $[a, b] \subseteq \mathbb{R}$ if $f(x)$ lies under (over) the line connecting $(a_1, f(a_1))$ and $(b_1, f(b_1))$ for all

$$a \le a_1 < x < b_1 \le b.$$

Concave up and down functions are also called **convex** and **concave**, respectively.

If f is concave up on an interval $[a, b]$ and $\lambda_1, \lambda_2, \ldots, \lambda_n$ are nonnegative numbers with sum equal to 1, then

$$\lambda_1 f(x_1) + \lambda_2 f(x_2) + \cdots + \lambda_n f(x_n) \ge f(\lambda_1 x_1 + \lambda_2 x_2 + \cdots + \lambda_n x_n)$$

for any $x_1, x_2, \ldots, x_n$ in the interval $[a, b]$. If the function is concave down, the inequality is reversed. This is **Jensen's inequality**.

Cyclic Sum

Let n be a positive integer. Given a function f of n variables, define the cyclic sum of variables $(x_1, x_2, \ldots, x_n)$ as

$$\sum_{\text{cyc}} f(x_1, x_2, \ldots, x_n) = f(x_1, x_2, \ldots, x_n) + f(x_2, x_3, \ldots, x_n, x_1) + \cdots + f(x_n, x_1, x_2, \ldots, x_{n-1}).$$

De Moivre's Formula

For any angle α and for any integer n,

$$(\cos\alpha + i \sin\alpha)^n = \cos n\alpha + i \sin n\alpha.$$

From this formula, we can easily derive the **expansion formulas** of $\sin n\alpha$ and $\cos n\alpha$ in terms of $\sin\alpha$ and $\cos\alpha$.

Euler's Formula (in Plane Geometry)

Let O and I be the circumcenter and incenter, respectively, of a triangle with circumradius R and inradius r. Then

$$|OI|^2 = R^2 - 2rR.$$

Excircles, or Escribed Circles

Given a triangle ABC, there are four circles tangent to the lines AB, BC, CA. One is the inscribed circle, which lies in the interior of the triangle. One lies on the opposite side of line BC from A, and is called the excircle (escribed circle) opposite A, and similarly for the other two sides. The excenter opposite A is the center of the excircle opposite A; it lies on the internal angle bisector of A and the external angle bisectors of B and C.

Excenters

See **Excircles.**

Extended Law of Sines

In a triangle ABC with circumradius equal to R,

$$\frac{|BC|}{\sin A} = \frac{|CA|}{\sin B} = \frac{|AB|}{\sin C} = 2R.$$

Gauss's Lemma

Let

$$p(x) = a_n x^n + a_{n-1}x^{n-1} + \cdots + a_a x + a_0$$

be a polynomial with integer coefficients. All the rational roots (if there are any) of $p(x)$ can be written in the reduced form $\frac{m}{n}$, where m and n are divisors of a_0 and a_n, respectively.

Gergonne Point

If the incircle of triangle ABC touches sides AB, BC, and CA at F, D, and E, then lines AD, BE, and CF are concurrent, and the point of concurrency is called the Gergonne point of the triangle.

Heron's Formula

The area of a triangle ABC with sides a, b, c is equal to

$$[ABC] = \sqrt{s(s-a)(s-b)(s-c)},$$

where $s = (a+b+c)/2$ is the semiperimeter of the triangle.

Homothety

A homothety (central similarity) is a transformation that fixes one point O (its center) and maps each point P to a point P' for which O, P, P' are collinear and the ratio $|OP| : |OP'| = k$ is constant (k can be either positive or negative); k is called the **magnitude** of the homothety.

Homothetic Triangles

Two triangles ABC and DEF are homothetic if they have parallel sides. Suppose that $AB \parallel DE$, $BC \parallel EF$, and $CA \parallel FD$. Then lines AD, BE, and CF concur at a point X, as given by a special case of Desargues's theorem. Furthermore, some homothety centered at X maps triangle ABC onto triangle DEF.

Incenter

The center of an inscribed circle.

Incircle

An inscribed circle.

Jensen's Inequality

See **Convexity**.

Kite

A quadrilateral with its sides forming two pairs of congruent adjacent sides. A kite is symmetric with one of its diagonals. (If it is symmetric with both diagonals, it becomes a rhombus.) The two diagonals of a kite are perpendicular to each other. For example, if $ABCD$ is a quadrilateral with $|AB| = |AD|$ and $|CB| = |CD|$, then $ABCD$ is a kite, and it is symmetric with respect to the diagonal AC.

Lagrange's Interpolation Formula

Let $x_0, x_1, \ldots, x_n$ be distinct real numbers, and let $y_0, y_1, \ldots, y_n$ be arbitrary real numbers. Then there exists a unique polynomial $P(x)$ of degree at most n such that $P(x_i) = y_i$, $i = 0, 1, \ldots, n$. This polynomial is given by

$$P(x) = \sum_{i=0}^{n} \frac{y_i(x - x_0)\cdots(x - x_{i-1})(x - x_{i+1})\cdots(x - x_n)}{(x_i - x_0)\cdots(x_i - x_{i-1})(x_i - x_{i+1})\cdots(x_i - x_n)}.$$

Law of Cosines

In a triangle ABC,

$$|CA|^2 = |AB|^2 + |BC|^2 - 2|AB| \cdot |BC| \cos \angle ABC,$$

and analogous equations hold for $|AB|^2$ and $|BC|^2$.

Median formula

This is also called the **length of the median formula**. Let AM be a median in triangle ABC. Then

$$|AM|^2 = \frac{2|AB|^2 + 2|AC|^2 - |BC|^2}{4}.$$

Minimal Polynomial

We call a polynomial $p(x)$ with integer coefficients **irreducible** if $p(x)$ cannot be written as a product of two polynomials with integer coefficients neither of which is a constant. Suppose that the number α is a root of a polynomial $q(x)$ with integer coefficients. Among all polynomials with integer coefficients with leading coefficient 1 (i.e., monic polynomials with integer coefficients) that have α as a root, there is one of smallest degree. This polynomial is the **minimal polynomial** of α. Let $p(x)$ denote this polynomial. Then $p(x)$ is irreducible, and for any other polynomial $q(x)$ with integer coefficients such that $q(\alpha) = 0$, the polynomial $p(x)$ divides $q(x)$; that is, $q(x) = p(x)h(x)$ for some polynomial $h(x)$ with integer coefficients.

Orthocenter of a Triangle

The point of intersection of the altitudes.

Periodic Function

A function $f(x)$ is periodic with period $T > 0$ if T is the smallest positive real number for which

$$f(x+T) = f(x)$$

for all x.

Pigeonhole Principle

If n objects are distributed among $k < n$ boxes, some box contains at least two objects.

Power Mean Inequality

Let $a_1, a_2, \ldots, a_n$ be any positive numbers for which $a_1 + a_2 + \cdots + a_n = 1$. For positive numbers $x_1, x_2, \ldots, x_n$ we define

$$\begin{aligned} M_{-\infty} &= \min\{x_1, x_2, \ldots, x_k\}, \\ M_{\infty} &= \max\{x_1, x_2, \ldots, x_k\}, \\ M_0 &= x_1^{a_1} x_2^{a_2} \cdots x_n^{a_n}, \\ M_t &= \left(a_1 x_1^t + a_2 x_2^t + \cdots + a_k x_k^t\right)^{1/t}, \end{aligned}$$

where t is a nonzero real number. Then

$$M_{-\infty} \le M_s \le M_t \le M_{\infty}$$

for $s \le t$.

Rearrangement Inequality

Let $a_1 \le a_2 \le \cdots \le a_n$; $b_1 \le b_2 \le \cdots \le b_n$ be real numbers, and let $c_1, c_2, \ldots, c_n$ be any permutations of $b_1 \le b_2 \le \cdots \le b_n$. Then

$$\begin{aligned} a_1 b_n + a_2 b_{n-1} + \cdots + a_n b_1 &\le a_1 c_1 + a_2 c_2 + \cdots + a_n c_n \\ &\le a_1 b_1 + a_2 b_2 + \cdots + a_n b_n, \end{aligned}$$

with equality if and only if $a_1 = a_2 = \cdots = a_n$ or $b_1 = b_2 = \cdots = b_n$.

Root Mean Square–Arithmetic Mean Inequality

For positive numbers $x_1, x_2, \ldots, x_n$,

$$\sqrt{\frac{x_1^2 + x_2^2 + \cdots + x_k^2}{n}} \ge \frac{x_1 + x_2 + \cdots + x_k}{n}.$$

The inequality is a special case of the **power mean inequality**.

Schur's Inequality

Let x, y, z be nonnegative real numbers. Then for any $r > 0$,

$$x^r(x-y)(x-z) + y^r(y-z)(y-x) + z^r(z-x)(z-y) \geq 0.$$

Equality holds if and only if $x = y = z$ or if two of x, y, z are equal and the third is equal to 0.

The proof of the inequality is rather simple. Because the inequality is symmetric in the three variables, we may assume without loss of generality that $x \geq y \geq z$. Then the given inequality may be rewritten as

$$(x-y)\left[x^r(x-z) - y^r(y-z)\right] + z^r(x-z)(y-z) \geq 0,$$

and every term on the left-hand side is clearly nonnegative. The first term is positive if $x > y$, so equality requires $x = y$, as well as $z^r(x-z)(y-z) = 0$, which gives either $x = y = z$ or $z = 0$.

Sector

The region enclosed by a circle and two radii of the circle.

Stewart's Theorem

In a triangle ABC with cevian AD, write $a = |BC|$, $b = |CA|$, $c = |AB|$, $m = |BD|$, $n = |DC|$, and $d = |AD|$. Then

$$d^2a + man = c^2n + b^2m.$$

This formula can be used to express the lengths of the altitudes and angle bisectors of a triangle in terms of its side lengths.

Trigonometric Identities

$$\sin^2 a + \cos^2 a = 1,$$
$$1 + \cot^2 a = \csc^2 a,$$
$$\tan^2 x + 1 = \sec^2 x.$$

Addition and Subtraction Formulas:

$$\begin{aligned}
\sin(a \pm b) &= \sin a \cos b \pm \cos a \sin b,\\
\cos(a \pm b) &= \cos a \cos b \mp \sin a \sin b,\\
\tan(a \pm b) &= \frac{\tan a \pm \tan b}{1 \mp \tan a \tan b},\\
\cot(a \pm b) &= \frac{\cot a \cot b \mp 1}{\cot a \pm \cot b}.
\end{aligned}$$

Double-Angle Formulas:

$$\begin{aligned}
\sin 2a &= 2 \sin a \cos a = \frac{2\tan a}{1+\tan^2 a},\\
\cos 2a &= 2\cos^2 a - 1 = 1 - 2\sin^2 a = \frac{1-\tan^2 a}{1+\tan^2 a},\\
\tan 2a &= \frac{2\tan a}{1-\tan^2 a};\\
\cot 2a &= \frac{\cot^2 a - 1}{2\cot a}.
\end{aligned}$$

Triple-Angle Formulas:

$$\begin{aligned}
\sin 3a &= 3\sin a - 4\sin^3 a,\\
\cos 3a &= 4\cos^3 a - 3\cos a,\\
\tan 3a &= \frac{3\tan a - \tan^3 a}{1 - 3\tan^2 a}.
\end{aligned}$$

Half-Angle Formulas:

$$\begin{aligned}
\sin^2 \frac{a}{2} &= \frac{1-\cos a}{2},\\
\cos^2 \frac{a}{2} &= \frac{1+\cos a}{2},\\
\tan \frac{a}{2} &= \frac{1-\cos a}{\sin a} = \frac{\sin a}{1+\cos a},\\
\cot \frac{a}{2} &= \frac{1+\cos a}{\sin a} = \frac{\sin a}{1-\cos a}.
\end{aligned}$$

Sum-to-Product Formulas:

$$\sin a + \sin b = 2\sin\frac{a+b}{2}\cos\frac{a-b}{2},$$
$$\cos a + \cos b = 2\cos\frac{a+b}{2}\cos\frac{a-b}{2},$$
$$\tan a + \tan b = \frac{\sin(a+b)}{\cos a\cos b}.$$

Difference-to-Product Formulas:

$$\sin a - \sin b = 2\sin\frac{a-b}{2}\cos\frac{a+b}{2},$$
$$\cos a - \cos b = -2\sin\frac{a-b}{2}\sin\frac{a+b}{2},$$
$$\tan a - \tan b = \frac{\sin(a-b)}{\cos a\cos b}.$$

Product-to-Sum Formulas:

$$2\sin a\cos b = \sin(a+b) + \sin(a-b),$$
$$2\cos a\cos b = \cos(a+b) + \cos(a-b),$$
$$2\sin a\sin b = -\cos(a+b) + \cos(a-b).$$

Expansion Formulas

$$\sin n\alpha = \binom{n}{1}\cos^{n-1}\alpha\sin\alpha - \binom{n}{3}\cos^{n-3}\alpha\sin^3\alpha + \binom{n}{5}\cos^{n-5}\alpha\sin^5\alpha - \cdots,$$
$$\cos n\alpha = \binom{n}{0}\cos^{n}\alpha - \binom{n}{2}\cos^{n-2}\alpha\sin^2\alpha + \binom{n}{4}\cos^{n-4}\alpha\sin^4\alpha - \cdots.$$

Viète's Theorem

Let $x_1, x_2, \ldots, x_n$ be the roots of polynomial

$$P(x) = a_n x^n + a_{n-1}x^{n-1} + \cdots + a_1 x + a_0,$$

where $a_n \neq 0$ and $a_0, a_1, \ldots, a_n \in \mathbb{C}$. Let s_k be the sum of the products of the x_i taken k at a time. Then

$$s_k = (-1)^k \frac{a_{n-k}}{a_n};$$

that is,

$$\begin{aligned} x_1 + x_2 + \cdots + x_n &= -\frac{a_{n-1}}{a_n}; \\ x_1x_2 + \cdots + x_ix_j + x_{n-1}x_n &= \frac{a_{n-2}}{a_n}; \\ &\vdots \\ x_1x_2 \cdots x_n &= (-1)^n \frac{a_0}{a_n}. \end{aligned}$$

Further Reading

1. Andreescu, T.; Feng, Z., *101 Problems in Algebra from the Training of the USA IMO Team*, Australian Mathematics Trust, 2001.
2. Andreescu, T.; Feng, Z., *102 Combinatorial Problems from the Training of the USA IMO Team*, Birkhäuser, 2002.
3. Andreescu, T.; Feng, Z., *USA and International Mathematical Olympiads 2003*, Mathematical Association of America, 2004.
4. Andreescu, T.; Feng, Z., *USA and International Mathematical Olympiads 2002*, Mathematical Association of America, 2003.
5. Andreescu, T.; Feng, Z., *USA and International Mathematical Olympiads 2001*, Mathematical Association of America, 2002.
6. Andreescu, T.; Feng, Z., *USA and International Mathematical Olympiads 2000*, Mathematical Association of America, 2001.
7. Andreescu, T.; Feng, Z.; Lee, G.; Loh, P., *Mathematical Olympiads: Problems and Solutions from around the World, 2001–2002*, Mathematical Association of America, 2004.
8. Andreescu, T.; Feng, Z.; Lee, G., *Mathematical Olympiads: Problems and Solutions from around the World, 2000–2001*, Mathematical Association of America, 2003.

9. Andreescu, T.; Feng, Z., *Mathematical Olympiads: Problems and Solutions from around the World, 1999–2000*, Mathematical Association of America, 2002.

10. Andreescu, T.; Feng, Z., *Mathematical Olympiads: Problems and Solutions from around the World, 1998–1999*, Mathematical Association of America, 2000.

11. Andreescu, T.; Kedlaya, K., *Mathematical Contests 1997–1998: Olympiad Problems from around the World, with Solutions*, American Mathematics Competitions, 1999.

12. Andreescu, T.; Kedlaya, K., *Mathematical Contests 1996–1997: Olympiad Problems from around the World, with Solutions*, American Mathematics Competitions, 1998.

13. Andreescu, T.; Kedlaya, K.; Zeitz, P., *Mathematical Contests 1995–1996: Olympiad Problems from around the World, with Solutions*, American Mathematics Competitions, 1997.

14. Andreescu, T.; Enescu, B., *Mathematical Olympiad Treasures*, Birkhäuser, 2003.

15. Andreescu, T.; Gelca, R., *Mathematical Olympiad Challenges*, Birkhäuser, 2000.

16. Andreescu, T.; Andrica, D., *360 Problems for Mathematical Contests*, GIL Publishing House, 2003.

17. Andreescu, T.; Andrica, D., *Complex Numbers from A to Z*, Birkhäuser, 2004.

18. Beckenbach, E. F.; Bellman, R., *An Introduction to Inequalities*, New Mathematical Library, Vol. 3, Mathematical Association of America, 1961.

19. Coxeter, H. S. M.; Greitzer, S. L., *Geometry Revisited*, New Mathematical Library, Vol. 19, Mathematical Association of America, 1967.

20. Coxeter, H. S. M., *Non-Euclidean Geometry*, The Mathematical Association of America, 1998.

21. Doob, M., *The Canadian Mathematical Olympiad 1969–1993*, University of Toronto Press, 1993.

22. Engel, A., *Problem-Solving Strategies*, Problem Books in Mathematics, Springer, 1998.

23. Fomin, D.; Kirichenko, A., *Leningrad Mathematical Olympiads 1987–1991*, MathPro Press, 1994.

24. Fomin, D.; Genkin, S.; Itenberg, I., *Mathematical Circles*, American Mathematical Society, 1996.

25. Graham, R. L.; Knuth, D. E.; Patashnik, O., *Concrete Mathematics*, Addison-Wesley, 1989.

26. Gillman, R., *A Friendly Mathematics Competition*, The Mathematical Association of America, 2003.

27. Greitzer, S. L., *International Mathematical Olympiads, 1959–1977*, New Mathematical Library, Vol. 27, Mathematical Association of America, 1978.

28. Holton, D., *Let's Solve Some Math Problems*, A Canadian Mathematics Competition Publication, 1993.

29. Kazarinoff, N. D., *Geometric Inequalities*, New Mathematical Library, Vol. 4, Random House, 1961.

30. Kedlaya, K; Poonen, B.; Vakil, R., *The William Lowell Putnam Mathematical Competition 1985–2000*, The Mathematical Association of America, 2002.

31. Klamkin, M., *International Mathematical Olympiads, 1978–1985*, New Mathematical Library, Vol. 31, Mathematical Association of America, 1986.

32. Klamkin, M., *USA Mathematical Olympiads, 1972–1986*, New Mathematical Library, Vol. 33, Mathematical Association of America, 1988.

33. Kürschák, J., *Hungarian Problem Book, volumes I & II*, New Mathematical Library, Vols. 11 & 12, Mathematical Association of America, 1967.

34. Kuczma, M., *144 Problems of the Austrian–Polish Mathematics Competition 1978–1993*, The Academic Distribution Center, 1994.

35. Kuczma, M., *International Mathematical Olympiads 1986–1999*, Mathematical Association of America, 2003.

36. Larson, L. C., *Problem-Solving Through Problems*, Springer-Verlag, 1983.

37. Lausch, H. *The Asian Pacific Mathematics Olympiad 1989–1993*, Australian Mathematics Trust, 1994.

38. Liu, A., *Chinese Mathematics Competitions and Olympiads 1981–1993*, Australian Mathematics Trust, 1998.

39. Liu, A., *Hungarian Problem Book III*, New Mathematical Library, Vol. 42, Mathematical Association of America, 2001.

40. Lozansky, E.; Rousseau, C. *Winning Solutions*, Springer, 1996.

41. Mitrinovic, D. S.; Pecaric, J. E.; Volonec, V. *Recent Advances in Geometric Inequalities*, Kluwer Academic Publisher, 1989.

42. Savchev, S.; Andreescu, T. *Mathematical Miniatures*, Anneli Lax New Mathematical Library, Vol. 43, Mathematical Association of America, 2002.

43. Sharygin, I. F., *Problems in Plane Geometry*, Mir, Moscow, 1988.

44. Sharygin, I. F., *Problems in Solid Geometry*, Mir, Moscow, 1986.

45. Shklarsky, D. O; Chentzov, N. N; Yaglom, I. M., *The USSR Olympiad Problem Book*, Freeman, 1962.

46. Slinko, A., *USSR Mathematical Olympiads 1989–1992*, Australian Mathematics Trust, 1997.

47. Szekely, G. J., *Contests in Higher Mathematics*, Springer-Verlag, 1996.

48. Taylor, P. J., *Tournament of Towns 1980–1984*, Australian Mathematics Trust, 1993.

49. Taylor, P. J., *Tournament of Towns 1984–1989*, Australian Mathematics Trust, 1992.

50. Taylor, P. J., *Tournament of Towns 1989–1993*, Australian Mathematics Trust, 1994.

51. Taylor, P. J.; Storozhev, A., *Tournament of Towns 1993–1997*, Australian Mathematics Trust, 1998.

52. Yaglom, I. M., *Geometric Transformations*, New Mathematical Library, Vol. 8, Random House, 1962.

53. Yaglom, I. M., *Geometric Transformations II*, New Mathematical Library, Vol. 21, Random House, 1968.

54. Yaglom, I. M., *Geometric Transformations III*, New Mathematical Library, Vol. 24, Random House, 1973.